T0074159

Never Split Tens!

Leslie M. Golden

Never Split Tens!

A Biographical Novel of Blackjack
Game Theorist Edward O. Thorp

Illustrations by the author

PLUS

Tips and Techniques
to Help You Win

C

Copernicus Books is a brand of Springer

Leslie M. Golden
Center for Computational Astrophysics
Oak Park, IL, USA

ISBN 978-3-319-63485-2   ISBN 978-3-319-63486-9   (eBook)
DOI 10.1007/978-3-319-63486-9

Library of Congress Control Number: 2017952373

Printed on acid-free paper

This Copernicus imprint is published by Springer Nature
The registered company is Springer International Publishing AG
The registered company address is: Gewerbestrasse 11, 6330 Cham, Switzerland

To my parents, Anne K. and Irving R. Golden, my beloved pets Duffy, Tweetie, Byron, Cicero, Skipper, Emerson, Newton, Maxwell, and Enrico, and Kitty, the feral cat who let me take her into my home, as well as Whitey, Blackie, and Momma, my most recent rescued felines, the family of birds in my window at my apartment on Broadway in Oakland, the fluffy white dog on Broadway near my apartment in Oakland, and Chirpie my robin and Finchie, the blinded red house finch I brought back to health and released, to a loud, exultant chirp of joy, and all the other needy, stray, and abandoned dogs and feral cats and kittens, opossum, raccoons, rabbits, squirrels, rats, mice, chipmunks, birds, and other wildlife, farm and zoo animals, fish, flying, burrowing, and crawling insects, and trees, bushes, plants, and grasses, all residents of the Earth, that I have be-friended, loved, saved, tried to save, and mourned. We don't own the Earth; we simply share it.

Oakley reasoned out loud, in furious thought.
"Blackjack, for crying out loud, is not a game of chance!"

Preface

This semi-biographical novel recounts how mathematician Dr. Edward O. Thorp revolutionized the casino industry with the development and popularization of card counting in his 1962 and 1966 editions of *Beat the Dealer*. That history stands by itself. This, however, is a novel, and so one may ask: what is the book about?

The apocryphal story goes that when the writers of *Fiddler on the Roof* approached producer Harold Prince, he asked them, "What is the play about?" To make the story short, they stammered and finally replied "tradition." He told them to go and write a song about tradition. Now, my book contains discussions of the meaning of life, including that of Somerset Maugham in *Of Human Bondage*, propagating the species, achieving pleasure, and providing food for other species, and I had let the characters conclude that the meaning of life was to gain immortality! We live so that we shall never die. Thorp has confided in friends that he is concerned that his contribution to probability theory will be recognized when he is no longer alive. Some in academia believe that Thorp's work in probability has made him an immortal. As Fiddler is about tradition, *Never Split Tens* is about immortality. By the way, they're the same thing, aren't they!?—cultural and individual immortality.

This book then is a semi-biographical, romantic novel about a mathematician who gains immortality. To do so, he must consort with gangsters to fund his gambling experiments in Nevada and Puerto Rico.

Because of the discussions of math in a biographical novel context, I consider this book to be "Stanislovski for card counters." Presenting theory in the context of narrative is a centuries-old grand literary tradition and style. In science, for example, we have Galileo's *Dialogues on the Two Chief World Systems* and *Dialogues on Two Sciences*. The influential acting books by Boleslavsky, Stanislovski, and, in modern times, Sanford Meisner, all present acting theory in the context of acting classes, a narrative. The mode appears in other genre.

Here, unlike the many books on blackjack which directly present the mathematics of the various systems in textbook style, I present many elements of the mathematics of card counting in the context of the narrative. I present some of the principles of probability that Thorp developed, some probability theory, calculation of indices, expectation values for the complete point count system, higher level systems, comparison of the ten count strategy and complete point count system, camouflage techniques, Small Martingale and Reverse Martingale, advantages and disadvantages of sitting in third base, and how a card counting system is developed. I also present money management and camouflage techniques. All are introduced within the context of the novel, a narrative.

I, in particular, frequently detail various camouflage techniques. These include ones developed by Thorp and others as well as ones I conjured up. These include coming into town to audition as a musical act or stand-up comic and getting toked to hit the tables; using the Bible as a guide to strategy, betting, and laying out when the deck is cold; using a rigged toy slot machine to do the same; playing a "perfumed" drunk there only for the free drinks; wearing a parka and ski boots to present oneself as a skier hitting the tables after the slopes have closed; and acting as a man desperately desiring to get invited to all those wild showgirl parties.

That relates to one of the reasons I wrote the screenplay, from which this book is reverse-adapted. From my years as a gambling columnist for five UK-based magazines, I learned that others had taken credit for concepts that Thorp had developed. For example, Thorp discussed the strategy of standing behind the players at a blackjack table, waiting for the deck to favor the player, and then sitting down and placing large bets. He called this "strike when the deck is hot," such an important strategy that I borrowed it for the title of Chapter 18. Other workers have coined terms for this, implying they developed the concept. He also introduced the concept of blackjack teams, referring to them as "cooperative friends." Others claim to have developed such a tactic. I find such behavior intellectually dishonest, and I wanted to somehow reclaim Thorp's inventions.

As I prepared this book, I realized that the gambling literature has not done justice to the contributions of those mathematicians who became known as the Four Horsemen of Aberdeen, Roger Baldwin, Wilbert Cantey, Herbert Maisel, and James McDermott, and to Harvey Dubner. The former group published the first basic analysis of blackjack, and Dubner electrified the standing-room only audience at the 1963 Fall Joint Computer Conference in Las Vegas with his presentation of the original "Hi-Lo" strategy. Based on conversations with Dr. Maisel, I discuss the former group briefly.

Based on numerous interactions with the Dubner family, I present a lengthy discussion of Mr. Dubner's professional life and contribution in an appendix. Because of Dubner's importance and that this may be the only "biography" of him ever written, I felt somewhat justified in displaying his son's passion for his father's position by quoting him at length. This sometimes leads to repetition of material.

I also include three other appendices in this book. One discusses the manner in which I altered historical fact for the purpose of the narrative, including a comparison of the dates I establish in the novel to the actual dates in Thorp's life. I note what fictional portions of the novel were created for the story itself. This appendix adds to the biographical value of this book. The second appendix presents a discussion of the concepts behind the twenty or so most well-known counting systems developed in response to the 1966 second edition of Thorp's *Beat the Dealer*. I list them in order of complexity rather than the puzzlingly non-instructive practice of all other gambling writers of listing them in alphabetical order.

Professional blackjack players know that skillful money management and effective camouflage in addition to facility with the mathematics of the card counting system are needed to ensure a winning career. Much of the pedagogical value of this book lies in the presentation of aspects of these two elements of card counting within the narrative. The acting aspect of camouflage has received little discussion in the literature of blackjack. As a professional actor, I add a third appendix, on using acting techniques as camouflage.

In short, I hope you'll find reading *Never Split Tens* to be an enjoyable means of obtaining many of the mathematical, money management, and camouflage skills needed to be a successful card counter. As I noted, you could consider it Stanislovski for card counters. On the other hand, you can just read it for the romance and its wit, charm, and humor. As my acting and writing mentor Del Close, the late guru of the Second City improvisation nightclub in Chicago would say, "All stories are love stories." *Never Split Tens* is, after all, a love story.

Introduction

How I Entered a Life of Sin and Degradation . . . and Playing Blackjack for Fun and Profit!

My twin brother Bruce and I had a gang of high school friends who, during high school, college, and young adulthood, would get together in our basement recreation/bar room for a game of poker, joke-telling, and pizza, pretzels, and beer. Bruce and I, Mike Flickinger (University of Illinois), Evan Jacobsen (Miami of Ohio), Dave Westerman (University of Michigan), Lee Campbell (Yale), Jack Camphouse (Naval Academy), Ira Epstein (University of Illinois), Dale Brozosky (University of Michigan), and sometimes Dick Haines (Miami of Ohio), Chris Jones (Occidental College), and Dennis Saliny (Purdue), with the occasional other visitor, would utilize our lofty intellects in playing for the high stakes of a nickel a chip, maximum three-chip raise, around that green fold-up circular pad we put on the card table. The chips were housed in a Quaker Oatmeal cylindrical box, decorated, as a Horace Mann elementary school project, with green wallpaper for our old marble collection.

If you lost $5, it was a devastating night. We became knowledgeable in the various types of poker games, such as In-Between, 5-card stud, 5-card draw,

7-card stud, Baseball, Night Baseball, Spit in the Ocean, High-Low, N-----, Indian poker, Hearts, and Chicago, as well as blackjack.

I'm happy to say that despite my bad influence, all of these friends have had successful professional lives. Only I entered into a life of sin and degradation.

It was only natural that when Bruce took a vacation to the Bahamas one year after college, he learned about the so-called basic strategy of playing blackjack. He told me about the Oswald Jacoby book that he had studied.

As a graduate student in astronomy at the University of California, Berkeley, I had been instrumental in founding the University of California Jazz Ensembles, the fabulous organization that has produced scores of professional musicians and the home of the Pacific Coast Collegiate Jazz Festival. In 1972, we went to our first jazz festival, the Reno Jazz Festival.

My brother and clarinetist Ron Svoboda, his friend in the Oak Park-River Forest High School orchestra, had formed The Deuces dance band. We bought the music from the previous band at Oak Park High, The Dominoes (whose graduates included trombonist/baritone singer Paul Kiesgan and trumpet player Bill Vogel of The Dukes of Dixieland); they had bought the music from the previous band at Oak Park High, A Band Called X-Squared. I also played trumpet in the Deuces. We were originally going to call the band "Deuces Wild" but decided on just "The Deuces," based on the name of the singing group, The Four Aces. The album we made, though, was called "Deuces Wild."

Among our illustrious alums is Jon Deak, the former assistant principal bassist with the New York Philharmonic Orchestra, who is renowned as a composer of classical music with jazz influence. Jon had been a piano player, who converted to bass for purposes of playing in the high school orchestra. He got his on-the-job instruction in jazz with the Deuces. My brother is proud to relate that he taught Jon how to play pizzacato (fingers only). Jon's father, an artist, painted the music stands with the Deuces playing card logo.

Bruce went to MIT for college. A high-note trumpet player, he was the lead trumpet player for the MIT Techtonians Jazz Band, under the direction of trumpet player Herb Pomeroy. Indeed, Herb, as Bruce related it, only took on the position when he heard Bruce play at his audition. At Cornell University, where I went to college, we regrettably had no jazz band. We had a group that rehearsed, to my recollection, once, in a room in the Willard Straight Hall student union.

When I went to Berkeley, I thought for sure that here in this hip place there would be a jazz band. There wasn't. Bandleader, composer/arranger, and trombonist Don Piestrup had had a group there in the 1950s, I was told, but when I arrived in 1966 no big band existed. I put an ad in the *Daily Californian* and got some replies. I remember that of the twenty or so who answered the ad, about twelve were guitar players. One fellow told me that although he had just started playing the guitar in a couple months he'd be ready to play jazz in the band.

The following year, one of the fellows who had answered my ad, Bob Docken, a trombone player in the Cal marching band, put his own ad in the Daily Cal. This time he had sufficient response and with the aid of Rick Penner, a trumpet player in the Cal band, the UC Stage Band started rehearsing in the Cal band rehearsal room on Sunday nights. I think the classic arrangements we played were from the Cal band library.

Membership and attendance was inconsistent, but we did put on a spring concert on Lower Sproul Plaza which was documented in an article I wrote for the Daily Cal. It included a famous quote from Bay Area jazz educator and KJAZ disc jockey, Herb Wong, "What, a jazz band at Cal!"

Only when Dr. David Tucker arrived at Berkeley in the fall of 1968 from the University of Illinois to be the arranger for the Cal band did the jazz group, the University of California Jazz Ensembles, or more commonly UC Jazz Ensembles, really gain stability. He had a strong jazz background and played regularly in the Tahoe show bands. Dave remained one of my two best lifelong friends, passing away in 2003.

Far from simply gaining stability, under Dr. Tucker's leadership, inspiration, winning personality, and energy, we soon had three big bands, numerous instrumental groups, and classes, and gave numerous performances. These included concerts every Thursday noontime on Lower Sproul Plaza and Friday night concerts in the Bear's Lair.

The band sponsored the fabulously successful Pacific Coast Collegiate Jazz Festival, later dropping the "Collegiate" part of the title. Over a weekend, scores of high school and college instrumental and vocal jazz groups from the western states competed. We had guest artists perform at the Saturday night formal concert with the premier Wednesday Night Band, including Ed Shaughnessy, Freddie Hubbard, Sonny Rollins, Hubert Laws, and George Duke. By 1972, Dave judged that we were ready to perform at our first jazz festival, the renowned Reno Jazz Festival.

This was my chance. I went to the famous Moe's Bookstore on Telegraph Avenue to look for the Oswald Jacoby book that my brother had read, and I discovered the 1966 edition of Edward Thorp's *Beat the Dealer*. Clearly, this was a far more advanced approach to playing blackjack than the basic strategy presented by Jacoby. I mastered the complete point count system and on our trip to Reno experienced my first winnings. We stayed at the memorable Bee Gay Motel.

My mastery resulted in part from practicing with my then girlfriend Diane Dring in the noisy Bear's Lair, to simulate the atmosphere of a casino. Diane's mother had been a cabaret vocalist and perhaps Vegas showgirl, if I remember. Was this an omen of things to come?

It was on this trip that Glenn Markoe's hitting blackjack on his one and only bet occurred. It was on this trip that I not only convinced myself that the system could be used to earn sizable cash but also made a believer out of a doubter. Glenn Markoe was a French horn player in the band, as well as the brother of comedian and writer Merrill Markoe, who was for many years the girlfriend of David Letterman as well as the head writer for "Late Night With David Letterman."

Glenn was a doubter. He doubted that a winning system could be devised to win consistently at blackjack. He doubted that I could master one of those systems. Worse, he doubted me when I told him the deck was hot, very hot.

A couple of guys in the band were walking from one large room to another in Harold's Club through basically a corridor with a couple tables slapped against the wall. One table had only one player, so I told Glenn and the other guys to stop. I watched the game progress. The deck started to get warm. Glenn wanted to move on. I said, "Hold on."

The deck got hotter. I whispered to Glenn, "The deck is getting hot. Hold on." The deck got very hot. I whispered to Glenn, "Put down $10." He looked at me, bewildered. I was a graduate student in astronomy. He was studying Egyptian tombs as an ancient art and archaeology graduate student. We obviously had a problem communicating.

I should note that, following other distinguished appointments, Dr. Glenn Markoe was Curator of Classical and Near Eastern Art and Art of Africa and the Americas at the Cincinnati Art Museum for twenty-three years.

The deck got even hotter. This was perhaps the hottest deck I had seen so far on the trip. I was nearly frantic as I whispered in Glenn's ear, "Put down the $10!"

Glenn still had that bewildered look on his face. The dealer was getting down to the end of the deck and I was afraid she was going to shuffle up and destroy our advantage. As a friend of the Greek scholar Socrates, I knew that only by Glenn laying down the bet and experiencing the euphoria (note, a Greek-origin word) would he really appreciate the beauty of the system.

One more time, "Glenn, bet the $10!" Glenn reached over the player and put the cash on an empty spot on the table.

Glenn was new to the tables and didn't wait for both cards to be dealt before peeking. An Ace. The dealer gave him his second card. A Jack. Glenn got blackjack! He got blackjack on his one and only bet on the trip. The dealer shuffled the deck. I said let's go. Glenn, predictably, gave me his bewildered look. "Why leave now?" In any case, Glenn was now a believer.

During the rest of my sojourn at Berkeley, I made monthly trips to Reno. I would leave on a Sunday afternoon, when the other casino players would be returning to the Bay Area, and stay through Tuesday afternoon. This schedule was selected to enable me to play alone at the tables, one-up against the dealer. It was in these trips that I began to develop the camouflage strategies that I relate in this book.

I knew I was on the right track when the "chicken soup" buses of San Francisco Chinatown residents were passing me going west as I was driving east to Reno. The chicken soup moniker comes from their bringing containers of soup with them as they left San Francisco for Reno.

One year in the 1970s, Bruce came out to California and we spent a month travelling through Nevada, hitting Tahoe, Reno, Carson City, and Las Vegas. He had by now read *Beat the Dealer* and employed the "ten count strategy" technique. Both of us having engineering degrees, we were interested to compare the results of the two Thorp techniques. Events of that trip form the basis for several anecdotes in this book.

In particular, the toy "magic" slot machine episodes mentioned in Chapter 14 and Chapter 15 were based on fact. Usually, to avoid being seen on the same shift, my brother and I would play on opposite sides of the street downtown. One day I appeared at the Four Queens, I believe, and the dealers asked about the magic slot machine. I didn't know what they were talking about and said so. They persisted, however, and then I realized it was my brother they had seen using it. I said, "Oh, you must mean my twin brother." They didn't believe the story about my being a twin, so later that day we came in together. The dealers were so entertained by the toy slot machine and our actually being twins that they henceforth would shuffle up the deck when we asked them to do so. We'd do that when the deck became cold, that is, in favor of the house. When the deck was hot, we'd tell them not to shuffle up the deck and they would oblige.

With my advisors' research grants, my earnings as leader of the "Les Morris Quintet," a dance band I formed in which I used other members of the UC Jazz Ensembles, and my gambling winnings, I was perhaps the first graduate student in the history of the Earth to come out of grad school with more money than that with which I entered.

Moving back to Oak Park in 1980, I took a sabbatical from playing blackjack. In addition to teaching, I studied improvisation with the late guru Del Close of the Second City improvisation nightclub in Chicago and began a career as an actor. After retiring from teaching, with gambling casinos now existing throughout Illinois, I queried various newspapers about becoming a gambling columnist. I was eventually directed to Lyceum Publishing in London, which was embarking on a new glossy print gambling magazine, *Gambling.com* magazine. I would eventually write for five London-based gambling magazines mainly on blackjack, but also on craps and roulette.

*   *   *

At some period in this new career, I decided to write a screenplay based on Thorp's life. I was upset that many workers in the field had appropriated his techniques, to the extent of one actually having a particular technique named after himself. Thorp had also suggested team play at blackjack, now famous from numerous gambling teams and a movie.

This book is a reverse adaptation of that screenplay.

The path to writing *Never Split Tens* has experienced many turns, detours, and new highway construction. As a result, I am indebted to many people. I would willingly pay back those debts with my blackjack winnings, but these words will suffice.

It all began with cornet playing and the Deuces. I am indebted to my dad, Irving, who bought us our first cornets and then my Bach Stradivarius trumpet, my mom, Anne, who schlepped us to Mr. Ben Purdom's Suburban Music and then Jerry Cimera's (renowned trombone soloist with the John Philip Sousa band) house and Adolph "Bud" Herseth's (renowned principal trumpet with the Chicago Symphony Orchestra) house for lessons, and to my twin brother, who started up that dance band and who found that Oswald Jacoby book.

I am indebted to my dear parakeet, Tweetie, who helped me through my graduate school years. I am indebted to those who originally created my interest in theatre, English teacher Nina Grace Smith and English teacher and theatre director Knowles Cooke of Oak Park-River Forest High School. I am indebted to my late, dear, dear friend, Dr. David W. Tucker, for ensuring that the institution of the University of California Jazz Ensembles would thrive and taking the Wednesday Night Band of the UC Jazz Ensembles to Reno in 1972, without which jazz festival trip I would never have encountered *Beat the Dealer*. I am indebted to the late improv guru Del Close for teaching me how to write a scene.

Oak Park, IL, USA   Leslie M. Golden

Acknowledgments

I am grateful to those who have nurtured my creativity: Nina Grace Smith, Knowles Cooke, Benjamin Purdom, Jerold "Jerry" Cimera, Adolph "Bud" Herseth, Dr. David W. Tucker, Del Close, and Professor Ann Woodworth.

For their valuable comments concerning factual and technical aspects of the manuscript, I am indebted to Aaron Brown, Professor Stewart Ethier, Professor Edward O. Thorp, and the staff of Edward O. Thorp and Associates.

For providing first-person histories of the development of blackjack systems, I am indebted to Dr. Herbert Maisel, Harvey, Harriet, and Robert Dubner, and Bonnie Ritzenthaler (Mrs. Allan) Wilson.

For their advice concerning the dramatic content of the manuscript, I am indebted to Professor Joyce Porter, Professor James McConkey, my agent Sharon Kissane, director Ryan Firpo, and actors Lanny Lutz, Rick Plastina, Paul Porter, Mike Ward, and Lana Wood.

I am indebted to those who have carefully read the screenplay version of the book, professor and actress Joyce Porter, the late Professor Lowell and Professor Helen Manfull of Penn State University, my agent Sharon Kissane, Ryan Firpo, the director of *Bet, Raise, Fold*, and actors Lanny Lutz, Rick Plastina, Paul Porter, Mike Ward, and Lana Wood. I am indebted to the Goldwin Smith Professor Emeritus of English Literature at Cornell University, my former professor, James McConkey, for not only reading the screenplay version but also encouraging and stimulating me throughout its reverse adaptation into this novel. I am also indebted to those writers and editors in the London gambling publishing industry, Dave Bland, Chris Young, James Aviaz, Jan Young, Philip

Conneller, and James McKeown, for publishing my columns and thereby allowing me to make a living while the concept of *Never Split Tens* evolved.

I would also like to thank my editors at Springer, (Ms.) Maury Solomon, Harry J.J. Blom, Hannah Kaufman, and Elizabet Cabrera for their skillful and dedicated attention to the manuscript.

Contents

Part I

Doubling Down

1

Recollections

May, 1964
Harrah's Casino
South Lake Tahoe
Stateline, Nevada

Of the fifty or so patrons standing by the craps table, the only men who stared at tthe $10,000 in $500 and $1000 gambling chips that the enticing Rosette had laid down were the eunuchs. A sign noting "Reserved for Private Party" sat on the table. Rosette wore an alluringly clinging orange chiffon dress, which coordinated nicely with her orange-colored $1000 "pumpkin" chips. Maybe those were real pink pearls that dangled from her necklace.

The ca-chinking of slot machines competed with live music, multiple conversations, and the nearby roulette croupier call of "Rien ne va plus." With their senses of sight and hearing overloaded with these multiple inputs, the sensory-inebriated crowd at the craps table laughed and yelled encouragement as Rosette rolled the dice and implored the gods of the craps table.

"Bring it home to your sweet peach pie momma. Momma needs a few new pairs of shoes!" Rosette apparently liked posh footwear.

My dad was seated about thirty yards away, the only player at a blackjack table. Large stacks of $25 chips lay on the table in front of him. He was playing three hands at the same time, and next to each were $100 in $25 chips. A King of Spades and ten of Hearts, a Queen of Clubs and a Queen of Diamonds, and a Jack of Clubs and a ten of Clubs, each hand with a value of 20, good blackjack hands, stared upward at the ceiling.

The dealer seemed mildly interested. She had seen it all, no doubt, but maybe not quite this. Her frilly white blouse obscured some wrinkled neck

© Springer International Publishing AG 2017
L.M. Golden, *Never Split Tens!*, DOI 10.1007/978-3-319-63486-9_1

skin, the indication of either too much weight or too much sun. "Betty Lou" read her I.D. name tag. Her up-card, a four of Clubs, must have felt out of place in the presence of all that royalty. Her down card. .. Well, they'd soon all find out.

My dad, Edward Oakley Thorp, sat in the chair by the stacks of chips. The twenty or so people watching were hushed. The noise was as loud as a Tibetan monk's breathing while in the lotus position at sunrise.

Betty Lou held the deck of cards in her left hand. Oakley – my sisters and I usually called him by his middle name, his father's first name – stared at the Jack of Clubs and ten of Clubs. He deliberately slid the ten of Clubs to the right of the Jack of Clubs and took four $25 chips from one of the stacks of chips and placed them next to the ten of Clubs.

"Split," he announced to Betty Lou.

The crowd gasped. Dad now had four hands on the table. Was this the stupidest man they had ever seen or the cleverest? Betty Lou dealt a ten of Diamonds on top of the Jack of Clubs and a nine of Hearts on top of the ten of Clubs. He now had four pat hands. Maybe he was the cleverest man.

Betty Lou stared at Oakley. Maybe she was daring him to split again. Maybe she didn't care. I'd bet on the latter.

Oakley waved his hand over the cards, indicating he was standing pat with each of his four hands. No more splitting. Betty Lou flipped over her down card or, as gamblers refer to it, the "hole card," to reveal a nine of Hearts, for a total of 13 if you're counting along. She dealt herself a Jack of Spades.

"Dealer busts," she announced, although the applause of the crowd extolling the David-like victory of their hero signaled the redundancy of such words. She took sixteen $25 chips from her faux-gilded tray of chips and placed four next to each of Oakley's winning bets.

"It's customary to tip the dealer for continued good luck from the blackjack gods," she advised.

"See, that's the problem. I'm an atheist," Oakley countered. It was clear already that he and Rosette, assuming they would somehow meet, would never attain a satisfying spiritual connection.

My dad was never really a nerdy math type of guy. He looked like a nerdy math type of guy, but surviving as a young boy in a blue-collar neighborhood on the northwest side of Chicago meant you had to be, if not tough, at least a smart-aleck. Oakley was definitely a smart-aleck. At this time he was in his late 20s. He was tall and thin, not athletic or muscled. If he'd been an athlete they would have called him "lanky." That appearance and the ever-present black-rimmed glasses fought a continual battle with his personality, the smart-aleck personality. If he had been born in, say, New York, Brooklyn, the Bronx, any of those areas, no doubt he would have become one of the world's biggest nerds, just like my college roommate. But he wasn't, and I was proud of him. All of us were, me, my sisters, and my mom.

Raking in his winnings, Oakley barely noticed a casino manager sitting down next to him. His red tie hung loosely at the neck, lying over a white-on-white embroidered shirt, the logo "Harrahs" embroidered onto it. He didn't wear a name tag. The buxom woman accompanying him could have been Rosette's sister.

"That's a nice stack of chips," the manager observed.

Oakley stared at the woman.

"You should know," he said, being, yes, the smart-aleck.

"Splitting tens. Beginner's luck, I guess, huh?"

"Right."

"Tsk, tsk. Don't you know you don't never split tens?"

"Oh, you know, I forgot."

Sticking his face into Oakley's, the manager tilted his head. "I seem to reco'nize your face. You were on t.v. for somethin' or somethin', right?"

"Splitting tens."

I think Oakley should have quit while he was ahead. The manager without a name went to scratch his nose and knocked the pile of chips in front of Oakley to the floor.

"Oh, I'm so sorry. Please do allow me."

Oakley bent over to pick up the chips, and the manager dove beneath the chairs with him. He elbowed Oakley in the kidney. You could still hear the clamor of excitement coming from Rosette's craps table over Oakley's restrained groans. As in the wild, never let your predator know your level of distress. It's what Oakley learned in the neighborhood.

Pulling himself up, Oakley found two muscular security guards standing by the table. These were the kind of miscreants who really, really, wanted to be cops, only their cognitive skills didn't allow them to pass the exam. No problem existed with their psychological makeup. Not especially needing the reinforcements, the manager advised, "We don't want your action here, pal."

"Oh, and I thought you were just trying to pick my pocket."

The security guards didn't respond particularly affectionately to this wise crack. They grabbed Oakley by either arm and dragged him away from the table. Oakley struggled to keep his balance.

"Your chips'll be counted and the cash it'll be waiting for you at the cage."

"How thoughtful."

"We don't want you here. You got. .. body odor, right Jimmy?"

"Right," said one of the guards. "You offend my smelling bad."

"Hey, Jimmy, be nice."

He continued offering his brotherly advice to Oakley as they walked to the cashier's cage.

"But, a lot of our patrons, they complain about it. Next time you walk into this establishment, my esteemed associates. .. well, I don' know if I'll be able to control 'em, you know? Somebody might find you sometime lying face down somewhere on a cactus."

Three Months Earlier
CBS Television Studios
New York City

On the right side of the stage of the television show "To Tell The Truth," as seen from the live studio audience, three, well, nerdy-looking men sat. They all wore black-rimmed glasses. They all had greenish, brownish tweed jackets over a white shirt and a black tie. They all looked like college professors. Maybe they all were college professors, but at the beginning of the show I could only be sure that Oakley was one. He was at MIT at the time.

The genial host, beloved by all Americans for his quiet, respectful wit, was Bud Collyer. He always seemed to have a chuckle in his voice, a result no doubt of truly enjoying his job. His desk sat to the left of the professor-contestants and faced that of the four celebrity panelists. I always loved that term, "celebrity panelists." Why not just "celebrities" or "panelists" or, for that matter, "guest stars"? "Celebrity panelists" seemed to be a term embraced by all Americans.

Tom Poston, the comic actor, was one, sitting on the far left end of the panel as viewed by the contestants. Then there were Peggy Cass, the comic actress, Orson Bean, another comic actor, and, at the far right end, the glamorous Kitty Carlisle.

Kitty wore a dress that had a rigid vertical bodice, cantilevered out about six inches from her body. I'm sure her dress designer was also the costume designer for Ming the Merciless, as played by Charles Middleton in the classic Flash Gordon series of Oakley's boyhood. After the question and answer portion of the program was concluded and the panel voted, Oakley would shake hands with Ms. Carlisle first. Vivian, who was to become my mom, was there as well, mingling with the celebrities and Edward Thorp impersonators, her genuineness and always well-coiffured moderate-length blonde hair guaranteeing that she would be noticed among the celebrities.

Bud spoke to the camera.

"That's our distinguished celebrity panel, and now let's meet our special guests." The nerdy men spoke in turn.

"My name is Edward Thorp."

"My name is Edward Thorp."

"My name is Edward Thorp."

It was now time for Bud to provide the background information. Then the celebrity panelists would interrogate the three, trying to determine who was "the real Edward Thorp." At the end of the show, after the panelists voted, Bud would say, "Will the real Edward Thorp please stand up." Often, the guy who two or most of the panelists thought was the real special guest would start to rise from his seat, then sit own, and then the real surprise guest would stand up to audience applause. Sometimes they had two fake risings.

Sometimes, if I remember, the real guest would rise, then sit down, then rise again. It's amazing how you can create so many combinations of risings and sittings among three people. A celebrity of sorts in his own right, the real surprise guest would answer a few questions from the celebrity panel. On this show, Oakley would rise and stay risen. Two of the four panelists would correctly guess he was the real Edward Thorp.

First, Bud had to provide that aforementioned background.

"Edward Thorp is a professor of mathematics. At the age of seven, he was calculating in his head the number of seconds in a year. While at the Massachusetts Institute of Technology, he developed a technique called 'card counting' that has enabled him to win tens of thousands of dollars playing the game of blackjack in casinos from Las Vegas to Puerto Rico. His book on the subject, *Beat the Dealer*, made *The New York Times* best-seller list. It is currently the most frequently requested volume in the Las Vegas Public Library."

After this public library note, Oakley broke into a wide, obviously genuine grin, a beaming smile he possessed that could illuminate an entire room. It's surprising that not all the panelists would guess, on that evidence alone, that he was the real Edward Thorp.

"Peggy Cass, let's start the questioning with you."

April 16, 1955
Zellerbach Auditorium
UC Berkeley Campus

A forty-foot-long cloth sign, blue on gold, draped over the entrance to Zellerbach Auditorium, read "The Pacific Coast Collegiate Jazz Festival – Welcome to UC Berkeley." Oakley told me he loved playing at the Pacific Coast

Collegiate Jazz Festival, the PCCJF, at Berkeley. Zellerbach Auditorium stands on the west side of Lower Sproul Plaza. You walk down the stairs to the plaza, and it looms in front of you. Mr. Zellerbach must have been a heavy alumni donor.

All the student demonstrations, okay, riots, during the Free Speech Movement of a later decade, would be held up the stairs, to the east, toward the Berkeley hills, on Upper Sproul Plaza, or simply, as it was usually called, Sproul Plaza. Because it faced the administration building, Sproul Plaza, with the legendary Ludwig's Fountain, named after a famous frolicking dog, allowed frenzied students to walk straight to the stairs of the administration building for their sit-ins or love-ins or whatever "in" was on the activist menu that day.

Lower Sproul Plaza, on the other hand, was a football-field-sized area for throwing Frisbees and listening to the loosely structured bongo band on a Sunday afternoon. Oakley told me the interplay of rhythms was both mesmerizing and relaxing. Completely unstructured, without a solo, there were up to a dozen men, each playing a given rhythm over and over. They played, literally, and I don't mean figuratively, I mean literally, for hours. Once, a white guy with no idea of the protocol started playing a solo. Oakley told me the guy got corrected quickly. Nothing is quite like a bunch of guys with foot-high Afros skewering you visually – figuratively, not literally.

At intermissions of concerts, the audience would stream out onto the plaza. No problems with traffic or bus noise would greet them. They'd just look up at the sky and enjoy the stars and the Moon until the house lights would flash, announcing the end of intermission and the beginning of the second part of the concert.

It was about 10:30 p.m. this Saturday night, and the winners of the competitions had been announced. The host band, from UC Berkeley, had invited the UCLA band to perform with them for the final concert. They both sat on the stage. The guest soloist, drummer Ed Shaughnessy from L.A., was playing with the host band from Berkeley.

The music stands in front of the saxophone players from UCLA had "Bruin Big Band" written on them in a fancy script. Maybe it was Peinaud. The music stands in front of the saxophone players from the Berkeley band simply had "UC Jazz" written on them, in block letters. Maybe Helvetica. If you were a high school band member attending the concert you got the unsubtle message that UCLA values football over academics while Berkeley values academics over football. I guess Zellerbach appreciated that.

The auditorium was jammed. Dave LeFebvre was directing the Bruin big band, and Dr. David W. Tucker was directing the UC jazz band. An audience of not only the college musicians performing at the festival but also Berkeley students and Bay Area jazz fans filled the auditorium.

Oakley, at the time 22 years old, stood in front of the bands. He played sensuous trumpet riffs accompanying a woman vocalist. That was Vivian, who as I said was to become my mom. Oakley played a top-of-the-line trumpet, the Vincent Bach Stradivarius. Along with the Benge, it was the finest trumpet of the era. If you, of course, had a lot of money, and were a symphony trumpet player or big-time jazz player, you could get Roland Schilke in Chicago to make one of his hand-made silver Schilke trumpets for you.

The Bach served Oakley well enough. It had a nice dark sound. I remember he complained that the E at the top of the staff was a little flat, so he had to use the alternative first and second valve fingering to stay in tune.

Vivian, in her early 20s, about five feet three inches tall, was slim, and wore an alluring but sophisticated cocktail dress. Vivian possessed the aura of a society gal, one of the 500, invited to the Cotillions, the coming-out parties. She always carried herself with respect. I still don't know exactly why she went for a nerdy mathematician. I guess, no, I'm fairly sure, she loved his intellect. He kept her off-balance with his smart-aleck personality.

I learned in my dating years that after a while a lot of women get tired of being wined and dined by the shallow types. One of my girlfriends, after we moved back to southern California for Oakley's position at UC Irvine, told me why she was attracted to me. Deborah Elwood said to me, "Because you're a good man." Mom, Vivian, must have felt the same way about Oakley.

She sang the tune "I've Never Been in Love Before," accompanied by Oakley playing licks and the joint bands:

I've never been in love before,
Now all at once it's you,
It's you forever more.
I've never been in love before,
I thought my heart was safe,
I thought I knew the score.
But this is wine that's all too strange and strong.
I'm full of foolish song
And out my song must pour.
So please forgive this helpless haze I'm in,
I've never really been in love before."

She stared at Oakley as she sang. Everyone in the hall must have gotten the feeling the look had more content than to ensure they were on the same musical page.

Now it was Oakley's turn. He played eighteen bars, the first two stanzas, and then Vivian came back with the chorus.

But this is wine that's all too strange and strong.
I'm full of foolish song
And out my song must pour.
So please forgive this helpless haze I'm in.
I've really never been in love before.
I've never really been in love before.

When the tune was over, the audience applauded, as much for the music as for the admittedly syrupy sensuality, I'm sure. There were whistles and cheers, too. Oakley and Vivian continued staring at each other. I'd bet more than one person in the audience was thinking, "Get a room!" respectable society girl or not.

Les, the emcee, walked onto the stage, applauding.

"Let's hear it for the Bruin Big Band from UCLA, under the direction of David LeFebvre, and trumpeter Ed Thorp." He pointed to them.

"Yeah! And our own Wednesday Night Band from the University of California, Berkeley, under the direction of Dr. David W. Tucker. Yeah! Featuring Vivian, Vivian Artisen on vocals." Again, he pointed to them. "Let's hear it! Come on!"

Oakley and Vivian held hands, stared at each other, and smiled.

*   *   *

If you drive east on Ashby Avenue southeast of the UC Berkeley campus, past the Claremont Hotel, you'd enter the tunnel that would take you under the Berkeley hills to the communities of Concord and Walnut Creek. There, summer days often have temperatures in the 90s. Similarly, south of the Burlingame hills you'll find the temperatures in the south Bay, Santa Clara, San Jose, and Palo Alto, to be considerably warmer than in San Francisco. The weather in San Francisco, Oakland, El Cerrito, Richmond, Hayward, Berkeley, and other towns within those hills is dominated by the cool temperatures of the bay. In the morning and evenings you frequently have to wear a light jacket. It was typically chilly this Sunday morning at 8:00 a.m.

The guitar player with the UCLA band sat on somebody's trumpet case next to the bus waiting on Bancroft Avenue behind Eshleman Hall and alternately played riffs from "Joy Spring" and blew on his fingers for warmth. A couple of band members helped the drummer carry his set from Zellerbach to the bus. They loaded it into the belly of the bus and then stood around listening to the guitar player before climbing into the bus for the ride back to a warm Los Angeles.

Oakley walked to the bus carrying his trumpet case. Vivian held his free hand with one hand and carried a bottle with the other.

"Here's some orange juice for the ride. To replenish your energy."

"Excellent! Something to remember you by. Well, it was nice."

"Is that it, then? Just a one-night stand between starving student musicians?"

"I'm not starving." He sipped from the bottle. "Not now, anyway."

"A one-night stand then it is."

"Correct me if I'm wrong," said Oakley, "but Friday night and Saturday night. That makes it a two-night stand."

Vivian unclasped her hand from Oakley's and gestured as she sang a line from "These Foolish Things."

You came, you saw, you conquered me.

Oakley took another sip. "Yes, a libation worthy of Edward the Conqueror. Except, I think you said I'm the greatest trumpet player ever. You conquered me."

"So this is it," said Vivian beginning to be resigned.

"Depends. What do you mean by 'it'?"

"Just like a law student! If you weren't such a good kisser."

The other musicians had long ago climbed into the bus. The driver beeped for Oakley.

"Trained embouchure. Well, g'bye."

"Edward?!"

"I gotta go home now and train my embouchure."

"But!"

"Is seven o'clock okay with you?"

"What?"

"Seven o'clock. I'll drive up in my red T-Bird next Friday. I'll be by about seven o'clock. And just because you're so nice, I'm taking you to Lake Tahoe." My father had to be the smart-aleck.

"Oh, yeah. I mean, yes. Yes. Yes, Edward. Oh, you. .. you. . .."

The bus door closed and Oakley walked to a window seat. He and Vivian waved good-bye. He threw her a kiss.

"Oh, you!" she cried out and kissed him back.

2

A Shayna Madel

"And for God's sake, Julian, if she wants to let you kiss her, then kiss her," his mother advised.

"And if she wants to let you. .."

"Marcel!"

"The boy is twenty-five, Anne."

Most of the European Jews who settled in Chicago in the decades around the turn of the twentieth century lived in small apartments in the Maxwell Street area. After earning a little money they moved west to the Lawndale district, centered on Douglas Boulevard and its wide boulevard park. The elementary school was named Theodore Herzl, after the pioneering Zionist. Immigrants from each little town, or *shtetl*, in Eastern Europe had formed their own orthodox synagogue along Douglas Boulevard. To this day, those synagogues remain, now converted to churches or community centers as the Jewish families have long sincenmoved. Jews with some wealth opted instead for the Hyde Park neighborhood, its focus being the University of Chicago. They would always consider themselves superior to the west side Jews.

As the Lawndale district changed, the Jewish families there moved further west to the Austin neighborhood. By the 1960s and 1970s those neighborhoods also changed, and the Jews as well as other Europeans moved further west or to the northern Chicago suburbs. The Braun family settled in a modest home on Long Avenue, one block north of Madison Street with its kosher bakeries, grocery stores, and butchers.

Austin at that time was home to a mixture of European descendants. The Jews lived north of Washington Boulevard. The Italians lived south of Washington Boulevard. They would each keep their property immaculate so

© Springer International Publishing AG 2017
L.M. Golden, *Never Split Tens!*, DOI 10.1007/978-3-319-63486-9_2

that the other group wouldn't utter disparaging profanities, terms with which we're all familiar. Their children all went to Austin High School together.

Inevitably, fights between the two groups occurred. Although the Jewish kids idolized their professional wrestler, Ruffy Silverstein, if some *telener* did them wrong, they would go to the Jewish enforcer, Larry Rosenberg, a tough kid who was with them at Austin High. Julian Braun had never mixed with the tough kids. He had been and remained a young man of the books.

The living room of the Braun family home was modestly decorated with a sofa, love seat, lamp tables, and shelves of books. Jewish homes are depositories of books. The complete *Encyclopedia Judaica* sat on the top shelf. The small dining room connected to the living room. It had a breakfront and a solid oak dining room table surrounded with chairs upholstered in red cloth. Unless company was invited, the dining room table would always be covered with 1/4″ leather-covered cardboard padding, to protect the wood. Tonight the padding had been removed and the *Shabbos* table cloth, of white vinyl with drawings in gold of fruit and flowers, was placed on the table. For Sabbath dinner that night, the table was set with silverware and dishes, a vase of flowers, Sabbath candles, and a loaf of challah bread on a silver tray partially covered with a white cloth.

Julian Braun wore glasses. He sat on the living room sofa being lectured by his parents.

"I've got work to do. How long are the Redmans gonna stay?"

"*Bench licht*, dinner, chatting, *ver vese*, who knows? I made a huge pot of *chunt*, like Bobe used to make. And everything from soup to nuts," answered his mother.

"Maybe you'll like her, Julian," his father offered.

"The women at Hadassah all say she's a lovely girl, a *shayna madel*, with a pretty face, a *shayna punum*."

"Lovely girl," Julian's father concluded.

"Lovely girl," repeated Julian, unenthusiastically.

"And she's got a nice job with Metropolitan Life," encouraged his mother.

"Oh, great, she sells insurance."

"Actually, she's an actuary," said Anne.

"Actually, she's an actual actuary," countered Julian.

"She likes math, Julian. You work at IBM. You like math." Anne was trying to find some reason for hope.

"You could be the father of a whole family of math geniuses," his father added.

"*Foon dein moyl ain gots everen*. You know what that means? From your mouth to God's ears. Just give her a chance, Julian."

"'Chance is a word void of sense. Nothing can exist without a cause.' That's what Voltaire said."

"Voltaire. Schmoltaire. Was he Jewish? Huh?!" responded his mother.

"Well, the *ferstinkener* cause, Julian, is that your mother wants grandchildren."

"I'd like to have some little *pisher* to clap my hands with and play '*pache pache kicheloch*' and to look at and *kvell* over and say *ah leben af dein kop* to. You know what that means?"

"I love your head. A long life upon your head." Julian had probably heard the phrase at every cousin's club meeting. "Such a good boy."

"And we both want you to find a place of your own."

"That's not true, dear. You should live and be well, Julian, you can live here as long as you want, God willing. You love my *kreplach*, don't you, honey? It's not all doctored up."

Charitably, the doorbell rang heralding the arrival of the Redmans.

"I know. What's her name?"

"Nancy. Redman. What kind of a name is that for a Jewish girl? 'Nancy Braun.' Now that's a name."

Anne raised her arms above her head and began to sway back and forth. She sang.

"*Lo mir a-le i-ney-nem,*
lo mir a-le i-ney-nem,
trinken ah glass e-le va-ah-ah-ine
lo mir a-le i-ney-nem …"

"Enough of the wedding song, mom!"

"I'll get it," Marcel volunteered.

"Julian, why don't you open the door for the Redmans?"

Julian stood up from the sofa, as much in resignation as wanting relief from the double-barreled attacks.

"Okay. Should I be picking my nose or scratching my *tuchus*?"

It was obvious that Julian and my dad would eventually form a solid friendship, based not only on their intellects and love of mathematics, but their personalities.

*   *   *

Hours later the only evidence of a *Shabbos* dinner was the vase with its plastic flowers sitting alone in the center of the otherwise clean dining room table. Marcel and Anne sat in two of the upholstered chairs, Anne wearing a somewhat soiled kitchen apron, the dishes having been done. Marcel scratched at his chin stubble and yawned.

Two others, the Redman parents, slept in the living room, slumped on the living room sofa. It was nearly 1:00 a.m.

The decorations of Julian's bedroom were nouveau American Intellectual. His stamp collection, framed, hung on the wall. A card table stood in the very center of the room, the only significant furniture outside of a bed and a chest of drawers. A chess board with chess pieces sat on the card table. Many of the chess pieces lay on the table off to the side of the board. Julian and Nancy sat at the card table.

Nancy Redman, also in her mid-20s, had short-cut black hair. She was thin, not strikingly attractive, not voluptuously attractive, but attractive. She had no outstandingly bad physical attributes. She had those blue eyes that have passed through to modern Jews resulting from the adoption of Judaism in the Middle Ages by a western Europe-bound tribe of fair-skinned Mongolians.

"Trivial," judged Julian. "Just assign a probability to each of your opponent's possible moves and then a random number generator selects the move based on the pre-assigned skill level and the various probabilities."

"Random number generator. Based on a digit of pi?"

"Any irrational number would work, but it's hard to avoid repetition of the sequence. Linear generators are better. The next in the sequence is a linear combination of the previous in the sequence modulo an integer."

"But then the sequence would be limited in length to the value of the modulus, right?"

"Clearly. So I'm using several in parallel with different values for the coefficients and then I use a master to select among the parallel generators."

"That's great. Even the error that a player might make could be assigned a probability. All the moves don't have to be rationally based."

"That's an interesting point," Julian approved. Then he laughed. "Throw in the possibility of stupid moves."

They both laughed at the concept. Nancy moved a knight.

"See, now you've moved your knight, and we reinitialize all the variables and begin a new decision tree."

"Unless you have enough memory to have set up decision trees on all the possible moves from the beginning of the match."

"That's the beauty of the 704/709 computing machine I have at work. We finally have enough memory and speed." Julian wasn't prone to bragging. He was simply being factual. He continued in a Yiddish accent, lifting his finger to the sky as a learned sage might have done in ages past.

"Dere vill come ah time. .."

Nancy laughed, and Julian, relishing the response, joined her in laughter. This high-brow frivolity was interrupted by a jiggling of the bedroom door knob and then a knock on the bedroom door.

"It's one o'clock," exclaimed Anne, now in the role of Mrs. Braun, from beyond the locked door.

"Why is the door locked? The Redmans want to go home. *Oy gevalt.* What are you doing in there?"

3

A Date with Patsy Cline

If you want a gambling vacation and live in northern California, you go to Reno. If you want a romantic vacation, and like skiing as well as a little gambling, you go to South Lake Tahoe. Stateline, Nevada, is the town. Like Reno and Las Vegas, it has numerous marriage chapels and convenience stores.

Oakley had kept his promise and had taken Vivian to South Lake Tahoe. They stood in line at the cash register in the convenience store. I doubt they'd remember the name. As all convenience stores, a rack of cigarette boxes hung on the wall. The clerk's name tag read "C. Patel" The most common surname of sub-continent India, Mr. Patel must have, judging from the latest international population figures, 10 million or so namesakes.

Oakley held a half gallon of ice cream and a jar of peanut butter. A customer dressed in a red plaid flannel shirt, jeans, and cap, standard trucker uniform, stood in front of them. He was filling out a Nevada lottery ticket.

"Excuse, me. My wife is due in five months, and we don't want to have the baby delivered here," said Oakley.

"You will have to wait your turn," responded Patel.

Oakley went into professor mode. "You might as give me your dollar," he said to the trucker-type. "You're not going to win."

"Somebody's gotta win," said the trucker-type. "Driving the American economy," bemoaned Oakley sarcastically.

"If you're so smart, what are the odds of the Nevada Pick Five?"

"I don't know. I only know the odds are lousy and you're wasting your money."

"Oh, yeah. Says who?"

"You're wasting your money."

"Yeah, sure. You're so smart," he taunted. "So what are the odds, pal?"

© Springer International Publishing AG 2017

L.M. Golden, *Never Split Tens!*, DOI 10.1007/978-3-319-63486-9_3

Oakley had to take on the challenge. He was, of course, perfectly able to do so.

"Okay. Okay. You've got fifty-two numbers to choose from, right? How many ways can you pick five numbers from those fifty-two? It's fifty-two times fifty-one times fifty times forty-nine times forty-eight ways of doing that. Since the order doesn't matter, you have five factorial ways of choosing the winning numbers. So fifty-two times fifty-one times fifty times forty-nine times forty-eight divided by five factorial, which is one hundred and twenty, is about," he paused. "Two and a half million. Your chances of winning are about one in two and a half million. And my wife craves some ice cream and peanut butter."

"Somebody's gotta win."

"Actually, closer to one in two point six million." Oakley had had another second to think.

"I'll make you a deal. See the cigarette boxes? There are what, five rows, about twenty boxes in each row? I'll write down the location of one of the boxes and I'll bet you five dollars you can't pick the box I wrote down. I'll pay you one hundred dollars if you get it right."

Mr. Trucker laughed mockingly, like a guy who's too smart to be taken in by a shyster.

"There's got to be a couple hundred boxes. No way, pal!"

It was not lost on Oakley that five times twenty is one hundred. He began to wonder if all the members of this fellow's family had vestigial brains, but then stopped himself mid-thought as being unattractively judgmental.

"The chances are one in a hundred but the chances in the lottery are one in two and a half million. Two point six," he said instead.

"Somebody's gotta win."

Mr. C. Patel had had enough and wanted to avoid violence, I'm sure.

"Okay. Okay. What do you have there?"

"Ice cream. Butter pecan. She craves it."

"I thought you also wanted peanut butter."

"She changed her mind. You changed your mind, right, honey bun?"

"Uh, yeah," Vivian, her mind still digesting the numbers, agreed.

"Women!"

Their purchase made, Oakley and Vivian walked to his car in the parking lot. It was, as noted, a bright red Ford Thunderbird. She carried a bag with the ice cream.

"What was that about?"

Oakley's outburst had put him in a precarious situation. He had to think quickly to get out of this one.

"Just playing games. Besides, you don't look pregnant."

Vivian didn't buy it.

"No, no, no. I'm not talking about the silly pregnant wife thing. The numbers."

It's hard for a math genius to play dumb, but Oakley tried.

"Yeah?"

"How'd you do the numbers?"

"Just made 'em up."

Vivian still didn't buy it. Oakley was a lousy liar, anyway. Two of the celebrity panelists, after all, had guessed he was "the real Edward Thorp."

"No, you didn't. I saw your mind working. I could see the smoke coming out of your ears." My dad didn't realize that my mom had transcendent powers and particularly acute ocular ability. She may even have possessed tetrachromacy.

"Edward, how'd you do the numbers? Are you some kind of math genius or something?"

Oakley looked at the fender of the T-Bird, wet his finger, and rubbed a spot. That one particular tiny dent bothered him every time he approached the car. Now perhaps Vivian would change the subject. He opened the car door for her and nibbled on her ear.

"Oh, so it's ear envy, huh?" He handed her the car keys with a modest request, "Can you put on the jazz station?"

Oakley walked around the car. Vivian stuck the key into the ignition and turned on the radio. He hoped the interrogation was over as he slid into the driver's seat of: the T-Bird.

A miniature dart board hung from the rear view mirror. Unfortunately, the radio was not playing jazz, but, instead, his anathema, country music. Patsy Cline singing "Walkin' After Midnight" was playing loudly, too loudly, on the car radio. Oakley knew the probing wasn't over.

> I walk for miles along the highway.
> Well, that's just my way of sayin' I love you.
> I'm always walkin' after midnight, searchin' for you.

Oakley cringed. "I hate country music."

She tried to elicit the truth out of Oakley by playfully swirling her finger in his ear. It was, however, only a diversionary tactic, and she persisted in her merciless interrogation. Oakley might have been in a bare warehouse, strapped to a wooden chair with 200-watt light bulbs shining in his face so that he couldn't see the faces of those grilling him about his seditious activities.

"It'll be jazz, but first. .. you gotta tell me about the numbers thing."

Oakley listened to the radio. He was caught. What to do? He bobbed his head back and forth and tapped his fingers against the steering wheel. Patsy Cline continued her torture.

I stop to see a weepin' willow cryin' on his pillow.
Maybe he's cryin' for me.
And as the skies turn gloomy night winds whisper to me.
I'm lonesome as I can be.

Oakley swallowed a primal scream.
"Actually, this one's not that bad," he blurted out.

4

Linear Operators

Oakley had been advised by one or more professors as an undergraduate at Cornell that cramming for exams was deleterious to your performance. You'd do better if you just cleared your brain for an hour or even a half an hour before the exam, particularly the two- or three-hour-long final exams, if you just took a walk.

Oakley wasn't aware at that time of the two modes of processing of the brain. Working on an exam utilizes the left mode of processing, the sequential, analytical mode. This was true even if you're writing an essay for an English exam. You still put the words on the paper one by one, in sequence. The right mode of brain processing, in contrast, is global, synthetic. I hate to use the word "creative," but creativity seems to flow from right-brain processing. If Oakley, while he was an undergraduate at Cornell University, had been aware of the split-brain research being done by Roger Sperry now thirty miles away from him at Cal Tech he would certainly have embraced regularly taking a stroll before exams.

His senior year at Cornell, he gave it a shot on a professor's recommendation. He performed brilliantly on the exam and from then on he followed the same routine. His chosen route was from College Town, past the Law School and Engineering Quad, past Willard Straight Hall, and onto the Arts Quad. He told me that he did deviate it from it once, before a cultural anthropology class he took as an elective. He noted that so many facts had to be retained for each tribe, their tools, institutions, and religion, for a few, that cramming until the last minute seemed the only way to get an "A" in the class. He slept for only one hour the night before the final exam, and aced it, to use the phrase in vogue at that time at Cornell.

© Springer International Publishing AG 2017
L.M. Golden, *Never Split Tens!*, DOI 10.1007/978-3-319-63486-9_4

At UCLA he found a route that also worked. His path of choice was north on the southeast part of campus, parallel to Hilgard Avenue. Plenty of palm trees graced the campus, but this route passed by a pine grove and its curious squirrels.

Oakley was walking with Professor Angus Taylor, age 43, and Allan Wilson, age 31, a mathematics post-doctoral student and Oakley's best friend. Professor Taylor was to author a book in 1959 on calculus, *Calculus with Analytical Geometry*, that was to become a standard in colleges for decades. He was Oakley's Ph.D. advisor.

Allan shuffled a deck of cards with one hand. He often brought it with him to class if the class became boring. Allan himself was a bright fellow. All three carried notebooks as they strolled through campus from the Mathematical Sciences building to get an ice cream cone at the Terrace Food Court.

Oakley was bothered by the lack of an attractive name for the building where he spent most of his time. Whereas many buildings on campus were named after donors, the buildings housing engineering, math, and the sciences had generic names. "Geology," "Physics and Astronomy Building," and "Mathematics Sciences Building" were institutionally bland enough. You had to wonder, though, about the effect on the psychologies of engineering students whose studies were housed in, and I'm not making this up, "Engineering I," "Engineering IV," and "Engineering V." I don't know what happened to II and III. Oakley would reminisce about the Ivy-covered names at his alma mater, Goldwin Smith Hall, Morrill Hall, and Lincoln Hall. Even the buildings on the Engineering Quad had names: Phillips Hall, Carpenter Hall, where the *Cornell Engineer* student-run magazine had its offices, Kimball Hall, and so on.

It was a typical end of class tumult, with students and faculty everyone going from one class to another or to a library. The over-abundance of pulchritude would distract a blind monk. You know, the altruistic type, those who would volunteer to lead seeing people across the street.

Allan stopped shuffling the deck to offer a suggestion to Oakley.

"You gotta tell her sooner or later, Ed."

"What do the cards say?" asked Oakley.

Allan knew that a standard deck of 52 cards couldn't provide the answer.

"Dr. Taylor, do you happen to have a deck of tarot cards? I left mine at home," asked Allan.

"Ed, you should be proud of being a mathematician. It's the language of science. Maybe she'll get excited by it," said Professor Taylor. Oakley didn't think it was the best advice.

"Yeah. Honey, wanna get a look at my hypotenuse? Can I bisect your acute angle with my linear operator?" Professor Taylor, as all brilliant men, had an excellent sense of humor. You could joke with him and not feel awkward.

"That's funny. Why do mathematicians always have to prove themselves?" Oakley wondered out loud. Professor Taylor loved it.

"Now, that's an interesting existential concept."

They all laughed at the implications. Professor Taylor greatly enjoyed bantering with his students. They would soon become his colleagues, and very little compares with the camaraderie of bright academics gathered at holiday parties or in the evenings at their annual academic meetings. Truly bright people have a shine in their physiognomies, generated I think by their ultrasensitivity to stimuli. It's like a San Andreas fault earthquake. The first time you experience it, you don't know what it is, but after being told, "Oh, did you feel the earthquake?" you'll be aware of it forever and can become the instructor for the newcomers who hadn't yet experienced it.

So it is with the faces of the intellectually gifted. You can recognize the shine in their faces, but until you're told that the person is a brilliant astronomer or mathematician or book editor, you don't know its origin, just what it is. Henceforth, you're sensitized to, and aware of, the glow.

"He's right," agreed Allan. "Use math humor on her."

"Come on. She finds out I'm lying about being a law student and it's all over."

Professor Taylor felt a need to defend his profession.

"You know, you can get rich doing mathematics."

"No," countered Allan, "I proved the contrary for my master's thesis."

"Ah, that's right. Brilliant piece of work." They all laughed at Professor Taylor's affability. The laughing ceased for a moment as a particularly pulchritudinous student sauntered past.

"Excuse me, does math humor turn you on? Can I show you my linear operator?" asked Oakley. She glared at the three of them.

"Nerd."

This irritated my dad. He lost his sense of humor.

"I am not a nerd," cried out Oakley. Professor Taylor was mildly shocked at the change in Oakley.

"Hey!"

Allan had to explain. He'd seen this manifestation before. It was like a cloud prowling in the back of Oakley's mind. With the proper provocation, it would emerge and blot out the rational, brilliant mind of my father.

"It's a sore point. Some of his schoolmates used to call him a nerd."

At Oakley's manual urging, they sat down on a bench to observe the view. He had to broach the obvious subject.

"It's coming to a head. I'm going up to Berkeley again this weekend, and she's asking the wrong questions."

"Time for us, then, to go to Plan B," suggested Professor Taylor.

"What was Plan A?"

"The linear operator joke," Allan said in mock support.

"I think you and Allan should take your girlfriends to Lake Tahoe. Try out the blackjack system Baldwin and his group just published in the *Journal of the American Statistical Association*."

"Hey, want to go to Tahoe with me and some math friends this weekend?" Allan yelled after the woman student who had spurned Oakley. "We can talk about Baldwin's paper." He turned to Professor Taylor.

"Interesting stuff." Professor Sorgenfrey, their colleague, had in fact already had shown the paper to Oakley, who had shared it with Allan. Now back to academic mode, Oakley agreed.

"Very interesting stuff."

"She obviously likes your mind. Show her a good time. Then, after a nice dinner, tell her the facts of life," Professor Taylor continued.

"Uh, Dr. Taylor, I think she knows the facts of life," Oakley noted, divulging what no one would consider a state secret.

"No sex talk on campus!" Allan reminded them.

"Facts of life. You mean, that you can't get rich doing mathematics?" bemoaned Oakley.

They all paused in resignation. "Well, maybe you can," Oakley continued with some hope. Remember what Shakespeare's Julius Caesar said."

"*Et tu*, Brutus?" joked Allan.

"He said, 'There is a tide in the affairs of men, which, taken at the flood, leads on to fortune.'"

5

Chipped Beef on Pillows

January, 1956

I join all the others who think that "Barney's" is a stupid name for a casino. I mean, who is Barney? Sounds like a farmer to me. Maybe he was a farmer and then got tired of it and so he built a casino in some soybean field. No one names a casino after their first name. You have Harrah's and the Riverside Hotel. Name it Cal Neva, to attract residents of both California and Nevada. Calling it Neva Cal would have sounded too much like a food supplement, so they chose Cal Neva. Casino restaurant food has a bad enough reputation.

Then you have all the gambling-sounding names and desert-sounding names. The Four Queens. The Golden Nugget. The Mint. The Horseshoe. The Sands. The Dunes. The Stardust. The Wagon Wheel. The Westward Ho is a motel, but it's still a western sounding name. Yes, I know about Harold's casino in Reno, but that is Reno. I just can't understand naming a casino after Barney. I expect to find one named Billy Bob's or Bubba's. Maybe it's the letter "B" at fault.

In any case, although they took rooms at the more luxurious Harrah's, Barney's in Stateline, Nevada, on the shores of South Lake Tahoe, is where Oakley, Vivian, Allan, and his girlfriend Joan decided to try out Baldwin's system. The men who would later become known as the Four Horsemen of Aberdeen – Roger Baldwin, Wilbert Cantey, Herbert Maisel, and James McDermott – were mathematicians at the Aberdeen Proving Ground in Maryland and, using clunky electrical/mechanical calculators of the era, perhaps the Friden brand, they had laboriously devised a basic system for the game of blackjack. With the approval of their supervisor, Abe Karp, they had spent several thousand hours over a period of one and one-half years in 1952

© Springer International Publishing AG 2017
L.M. Golden, *Never Split Tens!*, DOI 10.1007/978-3-319-63486-9_5

and 1953 working evenings and weekends. Their professional credentials, as well as the intriguing possibility that their system could work, attracted Oakley, Allan (the one-handed card shuffler), as well as Professor Sorgenfrey and Angus Taylor. I would guess that Professor Taylor would have loved to have made the trip with them, but, future colleagues or not, one has to retain one's distance from students as a tenured professor.

The green felt-covered blackjack table, with the name "Barney's" printed on it in red ink, included a silvery money tray containing rows of chips, colored according to their value and each emblazoned, of course, with the name "Barney's." Next to it was a discarded pile of cards. The cards were house cards, which is my way of saying they had the name "Barney's" printed on them.

Phil, the pit boss, stood behind the dealer. In his 50s, his name tag read, "Phil." Joey, a dealer in her 40s, stood behind the table. Her name tag read, "Joey." She was the kind of woman, a bit weathered, more than a bit bored with life, probably single with two or three bratty kids living with her at the trailer camp, who rendered unsupportable any fantasy that the casino dealers had wild sex parties every night after work. Thankfully their shirts did not have "Barney's" stitched onto them.

Small piles of $1 chips sat on the players' side of the table in front of Allan, Joan, Vivian, and Oakley. Single $1 chips lay in the betting squares at the four players' positions.

Allan, Joan, Vivian, and Oakley sat at the table, clockwise with Allan to the left of the dealer. All wore ski vests. It was an attempt, although poor, at camouflage, because the clean, unbruised, unfrayed condition of the vests made it was obvious that they had never been worn at an actual ski slope.

Joey dealt two cards to the four players from her left to her right and then to herself, one at a time.

"Most of our players bet only one-dollar at a time," she said to reassure her heavy bettors.

From her 52-card deck, Joey dealt the eight of Clubs and two of Diamonds to Allan, the seven of Spades and seven of Hearts to Joan, the three of Diamonds and Jack of Spades to Vivian, and the eight of Hearts and three of Hearts to Oakley. Her up card was the King of Hearts.

"Good," said Allan. "I mean, we're feeling so silly."

"No problem. Would you like a card?"

"I have to get to twenty-one, right?" asked Allan.

"That's the game."

"Okay, I'll take a card."

Allan knew the game, of course. They were trying to act as if they were absolute rubes, perhaps friends of Barney, Billy Bob, or Bubba themselves.

Joey took a card from the discard pile and scratched the green felt of the Barney's table with its edge.

"Here, you scratch the table. That's the way you tell the dealer you want a card."

"Cool," said Allan. "I'll take a card. Oh, sorry."

He scratched the table with his finger.

"You idiot," said Oakley. "She meant with the card."

"Bimbo," seconded Joan.

"I don't feel too stupid," said the post-doctoral student in mathematics at UCLA.

Phil walked over and stood next to Joey. Maybe he could prevent her from getting a headache. Joey dealt Allan the seven of Diamonds. Allan scratched the table with the card.

"You've got seventeen. That's a pretty good hand," offered Phil.

"But I've got to get to twenty-one. Card please."

"Oh boy," exclaimed Joey. She dealt Allan the eight of Spades.

"Sorry. You're over twenty-one."

"You wouldn't know it," said Joan. Joan clearly was as good an actor, if not better, than Allan or Oakley.

"So I lose?" moaned Allan.

"Just be happy you're not playing more than a dollar," sympathized Phil.

Joey took the four cards from Allan's position and placed them in the discard pile. She took the $1 chip and put it into the silver-lined tray.

"Bimbo. I'm not going to take a card," said Joan.

"You've only got fourteen, Joan," noted Vivian.

A man, about 40, walked up behind Vivian and watched the game. He was slight of build, with sensuous full lips, and his black hair was combed back into a pompadour. He had piercing eyes. His whiskers showed he hadn't shaved in a couple days, and the black dirt under his fingernails showed he had spent a lot of time at the tables during that time.

"Yeah, but I don't want to go over twenty-one, like Mr. Genius here," replied Joan.

"Then, you wave your hand over the cards," lectured Phil.

"It's black magic," observed Oakley. "I knew it."

"You kids crack me up. No, that's the way you tell the dealer you don't want a card."

"There's got to be some kind of system to beat this game," said Allan, the naive one.

"Lots," answered Phil. "Lots. But, kids, when a lamb goes to the slaughter, the lamb might kill the butcher. But we always bet on the butcher."

"Got'cha," replied Allan.

"I'm a vegetarian," Vivian noted.

"Ma'am?" asked Joey. She wanted to know if Vivian wanted to take a card or stand.

"I'm going to do what Joan did," said Vivian. She waved her hand over the cards.

"You've got thirteen, Vivian. It's bad luck." Oakley was continuing in the black magic vein.

"So I should take a card?"

"What do you think, Phil?" Oakley was enjoying the role playing. He assumed that the moderate intellects of the dealers and pit bosses could be manipulated. Pit bosses, he was beginning to realize, don't do much except stand around looking knowledgeable and authoritative. Every once in a while they'd get on the phone at their station, ostensibly to call to casino management but more likely checking up on the latest sports scores or calling the wife.

"I can't advise you kids how to play, you know that," said Phil, authoritatively.

Vivian waved both of her hands over the table.

"Abracadabra. I'm not taking a card. Final."

Oakley scratched the table with his cards.

"Is this a great woman? I'll take a card. Can't go over twenty-one if I have only eleven."

"Finally, a real gambler," said Phil. At least he had the mental ability for sarcasm.

Joey delivered the Queen of Hearts to Oakley.

"Blackjack!" Oakley yelled. "Read 'em and weep, my friend!" Oakley was obviously really enjoying the role playing.

"Way to go, Edward," congratulated Vivian.

"See, he took a card," said Allan.

Joey, far from getting the headache that Phil feared, laughed at the antics of these obvious neophytes.

"No, kids. It's a blackjack only if you get twenty-one on the first two cards."

"Oh," said Oakley. "What about the flush?" I think Oakley by this time was playing too stupid.

"Flush?" asked Joey.

"Yeah. I've got three Hearts."

The 40-year-old gambler had seen enough.

"This is pathetic," he said.

"Sorry," said Joey. "This isn't poker."

It's interesting, Oakley would tell me, that years later the casinos developed hybrid blackjack/poker games in which bets on poker hands would be accepted as side bets. That time, however, was in the future.

Joey flipped over her down card to reveal the four of Diamonds.

"Fourteen," she said. She dealt herself the seven of Clubs.

"Oh my God." Joan's utterance was truthful.

"Twenty-one," said Joey.

Joey took the $1 chip from Joan and the $1 chip from Vivian. She placed them in her tray.

"Oh, Joey, Joey, Joey," said Joan. Even for experienced players, it's hard to dissociate the dealer from the cards dealt. Unless they are cheating dealers, so-called mechanics, the dealer has no effect whatsoever on what cards are dealt when and to whom. Yet, it's human nature, and Joan had trouble believing that such a nice person, as Joey seemed to be, could have pulled a seven of Clubs to reach 21, defeating them both.

"We should have taken a card, Joan," cried Vivian.

"Push," Joey said to Oakley. He, too, recall, had a total of twenty-one. "Push" is blackjack jargon for a "tie." No money is exchanged.

Joey scooped up all the cards remaining on the table and put them in her discard pile. The 40-year-old gambler type sat down in the seat to the left of Oakley. He laid a one hundred dollar bill on his position on the table, neatly, in the center of the square where bets are supposed to be placed.

"Chips?" Joey asked.

"No," said the 40-year-old gambler.

"Money plays," exclaimed Joey. This was one time that the pit boss served a purpose.

Phil heard this, turned his head, and nodded that he was aware that the gambler was actually betting the entire amount rather than asking to buy chips.

"Guys, I don't think we're on the bunny hill anymore," said Vivian. It was questionable whether they had been on the bunny hill at all.

"I'd like to keep playing these stakes," said the 40-year-old gambler to Phil. The message could not have been clearer. Get these college punks off of the table and you'll get some real action.

"Mr. Marcum would like to play," said Phil. So, this little dark-haired man with the piercing eyes was really a well-known gambler. The pit boss not only knew his name but referred to him as Mister.

"You know," Phil said to Allan. "It's been enjoyable watching you and your friends gamble, but you think you might like some lunch?"

"Well, this is fun. We want to keep playing," said Allan.

Phil went over to his station and reached into a drawer. He pulled out four pieces of paper, walked back to the table, and handed them to Allan.

"Here," said Phil. "These are coupons for free meals in the restaurant. Why don't you use them now?" The message from Phil was also pretty clear. Get the hell off of my table!

"Do they have chipped beef?" You know what smart-aleck uttered this. Every college and university cafeteria has chipped beef on their weekly menu. No other restaurant in the civilized or uncivilized world has ever served it. Even Phil knew this.

"Sure," replied Phil. "And it's probably much better than what you get in the dorms."

"I think I'm really hungry, you know guys?" said Oakley. They had accomplished what they had set out to accomplish, and no need existed to continue the ruse.

*   *   *

The use of pleasant background music in the restaurant in Barney's, light, happy, joyous, wasn't accidental. Marketing employed it to prevent those who lost at the tables to lose their enthusiasm for gambling. No one was at the hostess' desk just inside the plastic miniature bougainvillea-decorated entrance to hear Allan, Joan, Vivian, and Oakley laughing hysterically.

"You guys were great," gasped Oakley. "Unbelievable. Really pulled the wool over."

"Did you see the look on the dealer's face when you said you had a flush?" inquired Vivian.

"I'm seriously thinking about becoming an actor and giving up law school."

"You are?" asked Allan.

A young girl, 18 years old or so, approached them, smiling. She wore a blue pastel dress, with a Barney's name tag with the name "Shelby" written on it. She fit the background music. She carried an arm full of menus.

"Four?"

"No. Three," said Allan. "He's just the restaurant reviewer."

I love that. The hostesses always ask the obvious question, and probably for the first time in her career she was perplexed. Perhaps she was even stunned.

"Oh, really," Shelby said. "Let me get the manager." She walked away.

"Restaurant reviewer?" asked Vivian. She was obviously as perplexed as Shelby.

"Vivian, this is the way we get the most expensive stuff on the menu for free," noted Allan. "Then we use the coupons for dinner."

"But I wanted the chipped beef," noted Joan, hopefully not telling the truth.

"That's probably the most expensive thing on the menu," observed Oakley to everyone's laughter.

College students, even graduate students, face economic challenges. I'm sure Oakley didn't do this, but he told me stories that as soon as you move out of the dorm into an apartment, the first thing you do is go to the university cafeteria to acquire silverware. Allan's ruse implied that he had not been above such shopping strategies.

Thad, his name judging from his Barney's name tag, approached with Shelby. His suit indicated that he was the aforementioned manager.

"Good afternoon, ladies, gentlemen," he said. "What paper or magazine are you with?" He undoubtedly knew from looking at them that the members of the group were college students. They responded in near unison.

"The *UCLA Bruin*," said Oakley.

"The *Berkeley Gazette*," said Vivian.

"*The Journal of Interior Design*," said Joan. Where she got that one I have no idea.

"*Better Homes and Gardens?*" tried Allan.

"Nice try, kids, but restaurant reviewers never identify themselves" was Thad's decision.

"So much for acting," said Oakley.

"But you kids tried, and I'm sure you don't have a lot of cash. So, here . . ." Thad reached into his pocket and pulled out four pieces of paper.

"These are free dinner coupons. You can use them tonight. Just remember us when you get rich and famous." Thad was a good guy.

"But mathematicians don't get . . ." replied Allan.

Oakley coughed loudly to drown out Allan. "Thank you, thank you, thank you so much, Thad." He peeked at Vivian to see if she had heard Allan's comment.

*   *   *

They had some sandwiches, chipped beef not being on the menu, no doubt to everyone's disappointment. They were having ice cream sundaes. Vivian had asked for extra whipped cream and maraschino cherries, perhaps an unsubtle message to Oakley.

Oakley told me that he loved Vivian's name. Not only did he love her first name, he loved her last name, Artisen. She was a creative person by any measure. An English major at UC Berkeley, Cal, as she called it, she also designed fashion and was an accomplished jazz vocalist. Her name, Artisen, although not spelled "Artisan," was nearly a perfect match to her life skills. Her parents were Hungarian Jewish immigrants, her father a businessman and her mother a homemaker. They encouraged her creative leanings, but she didn't think her talents were inherited, the only musical talent evident in the Artisen household being her father's harmonica playing. He particularly liked the tune "Jimmy Crack Corn," or "Blue Tail Fly." Vivian's favorite was "Begin the Beguine," by the famed Jewish clarinet player Artie Shaw.

She and Oakley, the mathematician, would permute the letters of her name to see what they would spell. Then they'd come up with stories about the word or phrase. The game had started early in their dating, when Oakley had scraped up enough cash to take her to a Frank Sinatra appearance at the Hollywood Bowl. They realized that the letters of "Sinatra" and "Artisen" were identical except for switching her "e" for one of his "a's."

With three vowels, and four of the most commonly appearing consonants in materials printed in the English alphabet – "n," "r," "s," – and "t" many words would result from the permutations. In fact, the most frequently appearing letters in the English alphabet are e-t-a-o-n-r-i-s-h. Of the first eight, Vivian's name only lacked the "o." Oakley, with enthusiastic Vivian's help, enjoyed playing the word game. They called it "The Vivian Name Game."

They came up with dozens. Some of their favorites were:

Senitar, provided on the application of someone from the hills of West Virginia for political office.

Air nets, missile defense system of Nationalicwz Militariczskicz Polski.

Air nets, (or . . .) National Security Agency installation in the atmosphere above Roswell, New Mexico.

Ranites, extinct tribe of bush people who were compulsive joggers.

Ai, stern, response of helmsman to Long John Silver's order, "Left and backward, you barnacle-infested tart scalawag!!"

In teras, how do we measure quantities greater than 2^{40}?

Is a tern, answer to the question, "Look! Is it an eagle, an albatross, a condor, or a pterodactyl?"

Ranties, a small cloth, in mint, chocolate, or bubble-gum flavors, to insert into the mouths of screaming, belligerent children.

Rise-Tan, a product which today would be banned as carcinogenic.

Ran SETI, astronomer Frank Drake.

Sir Ante, the nickname of the King of poker players.

Arsenit, the smallest amount that will make you ill.

Its near, after hiding in the basement, they sensed the approach of the alien.

Rise Nat, what Guinevere said to the newly knighted Sir Nat.

Sear tin, ethical disclosure the surgeon had to tell the Tin Man before his emergency appendectomy.

NSA rite, wasting a Commie pinko.

Entaris, the second brightest star in the constellation Virgo the Virgin.

An tries, plea of the mother of fifth grader An Hou Leong to his teacher after he failed another Relativistic Quantum Electrodynamics Field Theory exam.

Ra Nites, ancient, mystical Egyptian evening celebration honoring the Sun god.

Tin ears, the result of getting a transplant from a character in the "Wizard of Oz."

Wait. They came up with many more, including:

Ne a stir, Middle English for a quiet night.

E trains, one of the subway transit lines in New York City for which Billy Strayhorn did not write a jazz piece.

Sir Nate, the familiar name for Nathaniel, an obscure knight of the Round Table.

Sir Neat, America's best-selling closet organizer.

Tins Era, a recently discovered, yet to be studied, period between the Bronze Age and the Iron Age.

Insert A, the first words in Igor's instruction manual.

Near sit, the cause of a patient breaking his coccyx.

Ten sari, what you can get for a buck at the world-famous Chennai Flea Market.

Ire Ants, the most vicious type.

Ran Ties, a cheap Pakistani import.

Tri-Sane, a prescription drug for those with multiple personalities.

Irestan, the region in Ireland that the IMS (Irish-Muslim Separatists) want for their own.

Raisent, the third-person past pluperfect subjunctive of "to raisin."

So many permutations! I can only imagine the laughter they shared while playing this game. I wrote down this list to show you that mom did indeed love Oakley and his mind and Oakley did indeed love mom. A few dozen more are presented at the end of the story. Oakley wouldn't have wanted the narrative to get cluttered.

"Tomorrow, at the crack of dawn, I play Baldwin's system," announced Oakley as they finished their ice cream sundaes. "They think we're just kids. I wrote the strategy down on a little card, and they'll think I'm just another nut with a doomed system. They won't suspect anything. I'll report at breakfast."

"How'd you find out about it?" asked Vivian.

"One of Edward's professors," said Allan. Vivian was starting to ask some uncomfortable questions.

"A law school professor?" Vivian was puzzled. How could a law school professor become knowledgeable about gambling systems?

"Uh, sure. He teaches . . . uh, criminal law," Oakley said, coming to Allan's rescue, who was trying to rescue Oakley. "Oh what a web we weave," you know.

"What?" said a puzzled Allan, not knowing into which part of the web Oakley was headed.

"Uh, yeah, sure. Criminal, sure. It's always a chance you take when you commit a crime, you know. A game of chance." Oakley's laugh did not hide the small beads of sweat forming just above his eyebrows.

"I guess," said Vivian. If Allan and Oakley were truly intelligent they must have known that Vivian was purposely giving them an easy out.

"Like we say, commit the crime and do the time," as Oakley continued with the masquerade.

"Edward, tonight's the night, right?" asked Allan, issuing an order more than a question. He had probably had enough of the charade.

"You mean, you guys . . . ," asked an incredulous Joan.

"No personal questions for my good friend," cautioned Allan.

"Thank you for respecting that, Allan." Oakley was relieved that the interrogation had changed its course.

"But, tonight is the night, right Edward?"

*   *   *

Before Oakley even thought about blackjack seriously, he knew intuitively that you don't sleep where you gamble. An old expression says you "don't _____ where you eat," meaning don't perform any unseemly acts aimed at, or especially with, fellow employees if you want to keep your job. Professional gamblers, unless they are severely ego-deprived, know that the less the house knows about you the better. If you're a big player and stay in their hotel, they'll get your credit card number and check your credit, see if you're a big spender, meaning in their terms, a high roller, determine if you like to have women in your room, and get a list of who you speak to on the phone. And that's just within an hour of checking in.

If you gamble at Barney's, in other words, don't stay in their rooms. That's why Oakley was giving Vivian a massage in their room at Harrah's. Allan and Joan were down the hall.

The room had two Queen-sized beds, a 21-inch RCA color television, whose multitude of dials intrigued my father, and a compact refrigerator in which they had placed a small bottle of cabernet sauvignon for a gentle nightcap. Vivian sat in a chair in front of a large oblong mirror with what appeared to be genuine oak sculpted frame.

In a demure pink nightgown and robe, Vivian combed her hair when the pleasure of Oakley's shoulder massage allowed. He wore a shirt, untucked, and jeans.

"Left. Down. Little more. Ahh. Yeah, yeah, yeah, yeah, yeah. Good, good, good. Ahh, yeah." Vivian was obviously enjoying the massage. She leaned her head back towards Oakley.

"Allan talked about 'is tonight the night?' What d'ya got planned, stranger?"

As an undergraduate at Cornell, Oakley had bought a Honda 50 for the ride to campus from his apartment. It was at the end of Glen Place, on the ground floor of a house close to the Cascadilla Creek gorge north of East Buffalo Street down the hill from College Town. He had never been a motorcycle kind of guy, but two of the fellows in his Tau Epsilon Phi fraternity, Dan Stern and Stern's lifetime pal Ron Schendel, both had Triumph scooters, and they saved valuable class commute time, otherwise wasted in walking from the campus to the distant fraternity house. Stern became a successful engineering businessman and mayor of Hermosa Beach, and Schendel became a successful engineering consultant. One day Schendel had given him a ride back to the fraternity house on his Triumph and Oakley was sold. It wasn't that dangerous after all.

One winter, Oakley had to lift his bike over a snow bank and wrenched his back. It never fully healed. At UCLA, he exacerbated the muscle spasms one day playing in a softball game. He was rounding third, took a step, and – wham! – the spasm hit. The next day the Bruin band had a concert, and the expansion of his chest and back while playing the trumpet made the spasms worse. The next morning he literally could not get out of bed. It took him an hour just to get up to go to the bathroom.

He spent the next two months getting massages at the campus clinic three times each week. He learned the technique. First, you apply hot towels. Then you spread over a lotion to reduce any friction. Then you use your index finger and middle finger to bore into the spasm, loosening it. Then the masseuse would search for another knot.

He was well prepared to give Vivian any kind of massage she needed. The stress of the gambling, forcing her to be part of the deception, had tightened her up.

"What d'ya got planned, stranger?" she repeated.

"Nothing."

"Well, what did Allan mean by 'tonight's the night?'"

"I don't know."

"Edward, what did Allan mean by 'tonight's the night?'" Vivian was not to be deterred. Either it was something significant in their relationship, or it was going to be kinky, and she wasn't quite sure if he had the skills for that or, if he did, if she was interested in participating, extra whipped cream and maraschino cherries notwithstanding. Mom was a classy gal.

"I don't know."

"Edward." Maybe kinky wouldn't be so bad.

Oakley knew it was time. A better time there would never be. It had to be done. If he reported to Allan that he hadn't confessed, Allan might spill the beans, and Oakley feared Vivian would never forgive him.

"Okay. . . .," he began, taking a deep breath. He continued as rapidly as he could, to minimize the pain. Perhaps, he hoped, Vivian wouldn't catch on.

"See, fifty-two times fifty-one times fifty times forty-nine times forty-eight is about fifty to the fifth. I know that five to the fourth is six hundred and twenty-five, so to get five to the fifth multiply by another factor of ten and divide by two. That gives about six thousand divided by two, or three thousand. Use scientific notation. That's three times ten to the third. Multiply that by the tens, ten to the fifth, and you have three times ten to the eighth. Five factorial is easy, it's six times twenty, or one hundred and twenty. About ten to the second. So three times ten to the eighth divided by ten to the second is three times ten to the sixth, or three million. And I reduced it by about fifteen percent because I underestimated the numerator and five factorial is one hundred and twenty and not one hundred."

"You did that in two seconds?"

Oakley was astonished at this reaction. Vivian showed no anger, apparently just admiration.

"I was slow, I know," said a relieved Oakley.

"So you are a math freak. A monster."

Oakley put up his hands, curling his fingers toward her. He lifted his head and bared his teeth. Vivian had never seen this before.

"Don't ever, ever call me a monster. Or I will eat your succulent flesh. Bwwaaaahhhhh!!"

"Ten to the whatever. You're a math person. You're a grad student in math, aren't you?" Apparently she wasn't alarmed that her boyfriend might also be a ghoulish vampire.

"Yeah." There. If there was any doubt, the truth was out.

"The lawyer?"

"I lied."

Vivian rose from the vanity and moved to the bed. She slid under the sheets.

"Damn!" thought Oakley. "It's over. Now she'll tell me to sleep in the other bed. Damn!"

"Yeah, I knew that," Vivian said from under the sheets.

"Sure." Oakley really didn't know how to respond.

"The UCLA law school has a phone number. And although I'm just an English major I know how to use a phone. They taught us how in my senior seminar."

"Oh no! I didn't count on that!"

"I even found out the title of your doctoral dissertation."

"You did?"

"Something about 'linear operators.' I don't remember the rest because that term itself is such a turn-on."

"It is?!" Thankfully, it appeared the crisis had passed.

"So why did you lie?"

"Then, maybe not," Oakley reconsidered to himself.

Emboldened, or alarmed, he wasn't sure which, he sat on the side of her bed.

"I didn't want to lose you."

"I'm five foot three, one hundred and fifteen pounds. I'm not that easy to lose," Vivian noted.

"Ah, great," thought Oakley. "She's playing along. I'm safe. That was easy!" Now it was time to open up, show his vulnerability.

"Ever since I moved to L.A., I've found the women are only interested in you if you're a lawyer or a doctor. I faint when I get a shot, so I chose lawyer."

"Edward, I live in Berkeley."

"You do!?"

Vivian slammed a pillow into his face and pushed him down on the bed. She buried his face in the pillow. Oakley didn't fight, ability to breathe or not.

"I'm going to suffocate you unless you apologize."

"For what?" he muffled from under the pillow.

"For lying to me, you jerk."

"I apologize," still muffled. Vivian had no remorse for the murder she was contemplating. She kept the pillow over his face.

"Help! Help!" you could hear the poor muffled voice yell. "Call the constable! Please, someone call the constable! When Allan and Joan come for wine they'll have you arrested for murder. Help! Help!"

"They won't be here for another half an hour. Plenty of time for my getaway. I'll be in Dodge by dawn. I'm going to suffocate you unless you tell me you'll never lie to me again." At least, she had given him a way of avoiding death before he reached the age of thirty.

"I'll never lie to you again," the muffled voice said. "I promise to tell the truth."

"I'm going to suffocate you unless you tell me you love me." The stakes were getting higher. Oakley had to reconsider. He thought of one of the greatest laugh lines in radio history, when Jack Benny faced a robber, "Your money or your life!" Benny paused, and the robber repeated, "Look, bud! I said your money or your life!" After a short pause, Benny said, "I'm thinking it over."

Oakley wondered if he should tell Vivian he loves her or get suffocated. The muffled voice tried the sympathy ploy. "I can't breathe. I think I'm dying."

I wonder if my father, contemplating his premature demise, thought lovingly of his parents Josie and Oakley Glenn, or of the first kiss he may have received in kindergarten at the Dever elementary school in Chicago, or of his first elegant high school geometry proof, or if he simply regretted not having performed a background check on Vivian.

"I'm going to suffocate you unless you tell me you love me," Vivian repeated.

"Gurgle, gurgle, dying . . . end . . . is . . . near."

"I'm going to suffocate you unless you tell me you love me, Edward."

"I . . . love . . . gurgle gurgle . . . gasp . . . . ," said the muffled voice with apparently its last breath.

Vivian removed the pillow so she could clearly hear his final word.

"You."

"Yeah, I knew that," she said, throwing the pillow on the floor and pulling him towards her.

6

Graveyard Shift

Joey and Phil were working the graveyard shift at Barney's. Joey was at a table different from that of the previous day, but in the same pit. In South Lake Tahoe, some of the skiers/gamblers would gamble before the slopes opened. After sundown, unless you were skiing at a location with night lights, the skiers would go to their hotels or lodges, sit by the fire, have some hot cider or rum, and then take their shower. After dinner they'd hit the tables. Those were the skiers who would gamble for diversion. Those who hit the tables before the slopes opened were the gamblers who would also ski. The casinos are aware of the difference, and are accustomed to seeing full tables even before the graveyard shift was over, usually 7:00 a.m. It was now about 6:15 a.m.

Three players were at the table, along with Oakley. He sat to the right of Joey, in what is called "third base." That allowed him to see all the cards that were discarded by losing players before he had to make his playing strategy decisions. That should have been a tip-off to Phil that this was not the rube they believed they had met the previous day.

Judging from the piles of chips in front of them and the number of chips in their betting squares, none of the other players was a either a high roller or a big winner. Five $5 chips lay on the table in front of the player directly to Joey's left, in "first base," and four $5 chips lay in front of the player to his left. Each had bet two $5 chips. Four $1 chips lay in the betting square of the player directly to Oakley's right. He had no chips remaining other than those.

Oakley, in contrast, was playing only $1 chips. Fourteen $1 chips lay on the table in front of him, and he had placed only two $1 chips in his betting square. If he was trying to deflect any suspicion, his playing only $1 chips and

© Springer International Publishing AG 2017

L.M. Golden, *Never Split Tens!*, DOI 10.1007/978-3-319-63486-9_6

his low betting would probably do it. The time had not yet come when someone suspected of being a player using a system would be ejected, no matter the size of his bank and the bet. Again, Oakley intuitively must have known to keep his bets low.

Ten assorted cards lay face-up on the table in a jumbled pile.

The Queen of Hearts, three of Diamonds, and seven of Spades lay face-up on the table in front of Joey.

Joey scooped up the cards on the table and put them in the discard pile. She picked up all the chips that had been bet and inserted them into the money slot next to her on the blackjack table. Both she and Phil were somewhat surprised to see Oakley up in the morning. From his play the previous day, they would have guessed he was a skier who found gambling a diversion.

"Twenty," announced Joey. "Sorry, fellas."

The player next to Oakley rose up from the table.

"You're too tough for me, Joey," he congratulated her as he walked away, completely wiped out.

"Better luck next time," she encouraged him. We never really found out, but it seems to me that the more action a dealer generates the better chance she has of ever getting promoted to pit boss, or to the security room, if that would be considered a promotion.

"You're doing much better than yesterday," said Joey, pointing at Oakley's pile of chips.

"Bets," she said to all the players at the table, pointing her finger and rotating her hand from left to right.

The players sitting in the first two positions placed two $5 chips in their betting squares. Oakley placed two $1 chips in his betting square.

"I was up all night studying technique. I think I got a very satisfying system," answered Oakley. Joey looked at him, having no idea of the *double entendre*.

He lifted his left hand to display the index card on which he had written Baldwin's system. "See." Oakley's methodical, measured play and his continual referring to the card in his palm had attracted a small group of amused bystanders. Some smirked at his comment.

Beginning at her left, Joey dealt a card to each player and then herself, all face down. She then dealt a second card to each player, face down. She dealt the six of Clubs to herself, face up.

"We had a guy come in here with a system and he walked out with a thousand," countered Phil. "You know how? He came in with two." Oakley nodded agreement to the pit boss. This only confirmed his assessment that pit bosses did very little on the phone other than talk to their wives. This one, at least, seemed oblivious to any possibility of a system being developed for the game, the system, for example, of Baldwin and his colleagues that Oakley was testing.

"Yeah, systems players," Phil continued, smirking. "We love system players. We send cabs to pick 'em up."

The player in the first seat thought he had a pat hand.

"I stand," he told Joey. "The only system is the one named Lady Luck, my friend," he lectured Oakley.

Oakley would have expected a statement of such abysmal ignorance. By sitting in the first seat, the player prevented himself from seeing any cards that had been dealt and discarded from hands that had busted. He lost the opportunity to make the more informed playing strategy decisions that could be achieved by sitting further around the table.

The other remaining player, a man about Oakley's age but with the rough, reddened hands of a local farmer or farmhand hitting the tables before going to work, scratched the table with his two cards. Joey dealt him the ten of Hearts. He flipped over his two cards to reveal the nine of Diamonds and six of Diamonds.

"I'm almost wiped out," he twanged. "You're just too hot, Joey."

Young Old McDonald picked up his remaining two chips, stood up, stared at his chips sitting on the table, about to be scooped up, and left the table.

"Better luck next time," said Joey to the fellow, knowing he'd be back.

Oakley turned over his cards to reveal the nine of Spades and nine of Clubs. He placed a second pair of chips on top of the nine of Clubs.

"Split."

"Man, you got an eighteen," the first player raised his voice to Oakley. "You don't split up an eighteen."

"I still have a lot to learn," as he excused his poor playing strategy to the player. "Starting with the A, B, C's of it."

Staring at Oakley, Joey dealt the seven of Clubs next to Oakley's nine of Spades and five of Diamonds next to his nine of Clubs.

"See, man. You had a pat hand and you blew it away," observed the player to Joey's left, sitting at first base. "Now you got crap."

"Yeah. I guess I'd better stand on both."

Joey flipped over her down card to reveal the Jack of Clubs, giving her a total of 16. With less than 17, the rules of blackjack stated that she, as the dealer, had to take a hit. She dealt herself the seven of Spades, for a total of twenty-three.

"Dealer busts," she announced.

"Talk about good luck!" exclaimed Oakley.

"Yeah," said the player at first base. "Just don't do that again, man. You don't know squat about this game, man." He pointed to Oakley's index card. "And your system ain't worth a plugged nickel."

Oakley was slightly annoyed with this fellow and his use of jazz lingo in referring to him. If he had known anything about jazz he would have shown some recognition at Oakley's having quoted a lyric from the standard "Teach Me Tonight."

Joey flipped over the cards of the self-anointed expert player to reveal the Ace of Clubs and seven of Diamonds. He had a soft 18. She paid out two $1 chips to each of Oakley's hands and two $5 chips to the player with the soft 18.

"You both win," she said as she scooped up all the cards on the table and put them in the discard pile.

"You know," said Phil, who was observing the action. "I don't usually give advice, but eighteen is a pat hand."

"I never was very good in school," said the fellow who had earned the bachelor's, gaining election to Phi Beta Kappa, and master's degree in physics before embarking on Ph.D studies in mathematics. "Thanks for the tip," he added.

The first player moved all his seven $5 chips into his betting square while Oakley again moved two $1 chips into his betting square.

"Double or nothin'," the player announced. "What an outstanding betting move," Oakley said to himself. "This guy must be playing a very, very sophisticated system." Oakley looked at him with the disdain that only a mathematics genius could direct at someone he considered a registered rube.

Joey dealt a card to mister know-it-all, Oakley, and herself, all face down. She dealt a second card to each player, face down, and then she dealt the Ace of Diamonds, to herself, face up.

"Damn," said the player, ruing the moment he opted for his new betting strategy.

Joey lifted up one corner of her down card slightly, stared at it for a moment or two, and then flipped it over to reveal the King of Clubs. Oakley wondered about the motivation for the drama. Did she have a bad night and wanted to get revenge on the world or at least on two of its inhabitants by prolonging the torture by this imposed anticipation, or did she feel some sympathy for the two players?

"Sorry, fellas," said Joey.

"You're killing me, honey," said the know-it-all loser, with what Oakley was not surprised was a condescending attitude toward at least this woman. He rose from the table. "You're killing me."

Oakley thought, "The dealer has nothing to do with it, *man*."

*   *   *

Oakley, Vivian, Allan, and Joan sat at a table in Barney's restaurant for the breakfast appointment at which they would hear the tales of Oakley's gambling experiment. Four place settings were on the table, silverware, plates, and glasses of water. The non-stop pleasant music played. A waitress, who, judging from her name tag with the name "Janyce," would lead to even money saying she was the hostess Shelby's sister, stood at the table holding an order pad and pencil.

"And two scrambled eggs with hashed browns *au gratis*," Allan told the waitress.

"*Au gratin*," Vivian corrected him.

"*Au gratis*," he repeated.

"It worked once," explained Joan to Vivian. It was the silverware bandit trying for another handout. Oakley told me that Allan, despite the wealth he would accumulate later in life, would continue to play such games. Very bright people with a playful leaning employ such manipulation games, more for the neuron-to-neuron mental challenge with intellectual inferiors than any monetary gain.

"He stood on soft eighteen against a six," said Oakley, still considering the hands he had seen played. "Baldwin's tables say he should have doubled down."

"Did you want to order?" asked Janyce the waitress.

"These so-called good players have no knowledge of the game. They really think that the cards and the dice and the wheel get hot! They play based on ignorance and superstition."

"Like tapping the table with a chip for good luck," agreed Joan.

"Even we learned that trying not to go over twenty-one won't succeed," added Vivian.

"So you were the only gunslinger left standing, huh?" asked Allan. The poor waitress was being completely ignored. At this point she probably was thinking that these people were so unaware of her presence that they'd probably not give her a tip.

"I came in with $10 and I left with $18.50," reported Oakley in mock *braggadocio*.

"My hero," said Vivian, kissing him on the cheek.

"Baldwin's system works," was Oakley's final, official report.

"Did you want to order?" asked the waitress. She had probably not read the paper in the *Journal of the American Statistical Association*.

"Very interesting," noted Allan. "But guess who's picking up the tab."

"Sorry," Oakley said to the waitress. "For me, uh, also two scrambled eggs, orange juice, and a blueberry muffin. Easy on the amaretto-cinnamon-walnut hollandaise sauce supreme."

Janyce the waitress stared at him. *Maybe these guys were on drugs,* she may have been thinking. Oakley wondered if she had ever, or would ever, enjoy a hollandaise sauce. He hoped she would.

"Looks like this food thing is contagious," said Vivian.

"I was just a big winner at the tables," he explained to a bewildered Janyce.

The waitress walked away. It's a sad commentary, Oakley thought, that so many living in rural areas have little to do except hunt and procreate. Once they've procreated, they have plenty of time to consider names for the new baby. Shelby and Janyce apparently came from such a background. Their father probably never thought that the animal he had most recently killed may have been the father or mother of a child who was also joyfully anticipated and dearly loved.

"Enough gambling talk," pronounced Vivian. "I have an important announcement to make."

"Oh?" asked Joan.

"Edward told me last night he's not really a law student."

"Way to go, Edward," Allan congratulated him. "Finally."

"I don't know why he's so ashamed of being in med school." Oakley's jaw fell open and he stared at Vivian in disbelief and new-found admiration that she was going to continue the masquerade.

"Med school?" inquired Joan, staring at Oakley.

"That's right. And I feel so good finally getting it off my chest."

"I guess it must be a good feeling," said Allan. "Right, Edward?"

"Yep."

"Well, Edward," Joan thought she'd best be careful here. "Uh. What are you going to specialize in?"

"He told me he's going to be a surgeon, right, Edward?"

"Yep. A surgeon. Brain, heart, elbows, you name it."

"He's going to do operations," Vivian gleamed. "He's going to be the world's greatest linear operator." She laughed at her joke making. She was obviously pleased at having confused her friends.

"You bum," Allan said to Oakley. "I'm going to kill ya'."

"Is she the greatest? Is she or is she not the greatest?"

"That's why he proposed last night," explained Vivian.

"I did? My head must have been in a cloud."

"Professor of Mathematics and Mrs. Edward Oakley Thorp," Vivian mock imagined their being introduced to society.

"And I said 'Yes.' I'm so happy."

"With these guys you don't know what to believe," observed Joan.

"It's part of his penance for lying to me," further explained the bride-to-be.

"See," Oakley bemoaned his fate. "This is why men choose celibacy."

7

Royal Garden Blues

It was barely 10:00 a.m. when they walked out of Barney's. Although it was January, it was a temperate 48° in South Lake Tahoe. The Sun helped, and its rays shone into their eyes. The ski operators weren't complaining. The roads were clear, and the temperature dove as you ascended through the gradually thinning air into the mountains.

No slot machines stood on the sidewalk outside the door, but three "Your Money or Your Life" brand vending machines were planted there. Guaranteed 100% sugar candy of various colors were in one. Souvenir trinkets were in the second, and the ever-present peanuts in the shell with a surprise gift were in the third.

"Okay," said Oakley. "Check-out time is in two hours. If this is going to get done, we've got to work in teams."

"Get what done?" asked Allan.

"Our wedding," said Vivian. It's hard to tell if she was continuing the joke or getting peeved at Allan.

"You guys weren't kidding?" Joan came to the aid of her boyfriend.

"I said 'Yes,' " Oakley reminded them.

"How romantic," said Allan, still perhaps in disbelief.

"I didn't even have to get down on my knees," said Vivian.

"She suffocated me with love." If it's not obvious by now, my parents cared for each other and found joy in their relationship.

"What a sweet, lovely story for the children," noted Allan.

"Okay," said Oakley. He was serious and it was time to get down to the business of nuptials. "Allan and I. My tuxedo. And a wedding chapel. You two, a wedding dress."

© Springer International Publishing AG 2017
L.M. Golden, *Never Split Tens!*, DOI 10.1007/978-3-319-63486-9_7

"I'd like some flowers," Vivian purred.

"And a florist," Oakley added.

"Edward, we need rings," said Vivian.

"There's a problem with money."

"No ring!? You just won $8.50."

"I picked up the entire tab."

"We need rings," Vivian demanded.

Oakley stared at the vending machines.

"Okay," he relented. "Who's got change?"

Allan, the consummate soon-to-be-designated best man, dug in his pocket for change and came up with a few dollar bills.

"Couple dollars," he announced. Oakley opened up his wallet and gleefully displayed a few dimes and nickels.

"Bunch of nickels and dimes," he said. He walked over to the vending machines and stared at the one containing the trinkets.

"This one. Perfect."

He put a dime in the slot and pulled the lever. A large pink gumball rolled out.

"How about this?" he said to Vivian. "Pink. We could hollow it out." Oakley was suddenly interested in jewelry design.

"That's good with me," she said. "I want a divorce." Of course, she was kidding. You can't get a divorce if you're not yet married. On the other hand, if they had been married, perhaps this would have been grounds for dissolution.

Oakley put another dime in the slot and pulled the lever. Another large gumball rolled out. He looked at Vivian, who wagged her head "No." He put in another dime and pulled the lever. A key chain rolled out.

"Okay," yelled Oakley triumphantly. "That's one."

They all laughed at Oakley's mock triumph. Yet, the key chain would have to do. He placed a fourth dime in the slot and pulled the lever. A pair of fuzzy dice rolled out. They were colored red with white dots.

"Well," Allan said. "That's for good luck."

"What if they're loaded?" asked Oakley.

"I'm content with a single-ring ceremony, Edward," said Vivian. If they really were going to do the necessary groundwork, get married, and check out of Harrah's before noon, they couldn't spend any more time playing vending machine roulette.

"Okay with me," said Oakley, apparently resigned to having a single-ring ceremony.

"You're the best man," he said to Allan. This was not the greatest surprise.

"Okay with me," he agreed.

"Allan and I will take the car," Oakley directed. "You guys scour these streets. We meet back here in an hour. Soldiers, synchronize your watches. Did we forget anything?"

"The guest list, the bridal registry, the invitations, the menu, the centerpieces, the seating arrangements, the orchestra, music for the first dance, and rice," offered Vivian, knowing that none of those embellishments would be realized.

"Rice," said Oakley. "Right." Oakley, if you recall, was a smart-aleck.

He looked at his wristwatch. "Fifty-nine minutes."

*   *   *

Oakley decided that the nuptials would be held in the Royal Garden Wedding Chapel. He and Allan had passed a number of chapels, and one can be sure that they all offered the same spiritually fulfilling, righteous, respectful, and, most importantly, statute-satisfying wedding service, suitable for all faiths, although they all would only be about ten minutes long. The cost would be about the same at any of them.

As soon as Oakley saw the name, however, he announced his choice. "Royal Garden" didn't appeal to him because of any pretensions toward monarchs and their accoutrements and fawning serfs, but rather because Royal Gardens Blues was the first tune played by the high school dance band in which he had first played jazz. He could still scat sing the tune, "dah, da da da da dot dot dot, de dah dah, da da da da dot dot dot." A simple little blues ditty; it was perfect for a high school dance band just starting out, and Oakley found that it remained in his brain, waiting for a reprise at any particularly unoccupied moment.

The dance band was led by a bookish kid from the concert band who played the sax, and his concept of putting together a dance band for sock hops wasn't to get any gigs but to put on a concert. Every week on a Sunday afternoon, the guys would get together at his house and rehearse a couple tunes. He only had a couple of arrangements, or charts, as jazz players call them. Royal Garden Blues was the first tune they'd rehearse every Sunday. They had it down pat, but the band never got any gigs.

Oakley decided this nice bandleader was a lousy businessman, and so he formed his own band. He actually bought some charts and within weeks got booked not only for the sock hops in the basement of the First Baptist Church the evenings of Saturday afternoon football games but also a couple private parties and a wedding. Yes, The Deuces, composed of high school kids, got a booking for a wedding.

The bass player in the band would become a music major in college and get hired by the New York Philharmonic, with whom he would gain renown for not only his musicianship but also for his bringing jazz to the composition of orchestral pieces. His father was an artist and painted their music stands white with red and black sketches of two overlapping playing cards, the deuce of Hearts over the deuce of Spades. Oakley taught Jon how to play pizzicato, using the fingers to pluck the strings instead of the bow and Jon would use such an embellishment in his compositions.

Royal Garden, it was obvious, meant a lot to Oakley, and he decided instantaneously that he and Vivian would get married there, even it meant scat singing the tune in his head for the entire ride back to L.A.

Walking inside, the first thing that Oakley and Allan saw was the pulpit. Tastefully done, it was a six-inch platform surrounded by a white picket trellis on which white roses and green vine climbed. Actually, the white roses and green vine would have climbed there, but plastic does not grow, let alone climb. The second thing they saw was a sign bordered with plastic white roses reading "Love the Lord." The white roses must have signified virginity, they realized. The third thing they saw was a sign behind the pulpit reading "The Royal Garden Heavenly Harmony Chapel. $25. Cash only. No checks, money orders, or Green Stamps. Pastor Elmer, Clergyman." An organ stood next to the pulpit.

Oakley couldn't figure out the reason for the words "heavenly harmony" in the name. Maybe the pastor felt that the harmonies of the Royal Garden Blues were presented by none other than the Lord. Oakley made a mental note to tell the pastor that in fact the blues progression was the blues progression, and that the harmonies were made by Louie Armstrong, Billie Holiday, Charlie Parker, and the composers for Duke Ellington, Count Basie, and others.

A man, no doubt Pastor Elmer, and a woman stood in front of the pulpit and waved hello to Oakley and Allan. The good pastor wore cowboy boots.

"The wife Alice and I have presided over many marriages," he said in a mild country and western twang, gesturing to his wife. "But it's usually, well, between a man and a woman."

"As long as they love the Lord, Elmer dear," sang Alice, in her own country and western twang.

"Yes," he agreed.

"You *do* love the Lord?" inquired Alice with a steely yet affectionate glare. It was beyond her understanding that anyone could in fact not love the Lord.

*   *   *

At about the same time, a few hundred yards from Harrah's, Vivian and Joan had found a florist shop. They stood in front of a refrigerated flower case. Her hair teased in a high bouffant, a salesgirl, in her 30s, stood next to them. She wore black lipstick and large earrings. She chewed gum. Her name tag indicated her name to be Infinity.

"If you want a love that will last for eternity," Infinity said looking straight into their eyes, "I always suggest the lilies from Roswell, New Mexico."

She moved sufficiently close that she forced Vivian to retreat a step, transmitted her less than pleasing body scent into Vivian's nostrils, and allowed Vivian to examine her moustache. She whispered, the reason no doubt for the close encounter, looked from side to side to ensure that no one who might consider doing her harm could hear her, and opened her eyes wide.

"They were planted by the aliens."

*   *   *

A sign on the wall of the musky tailor shop read, "Eisenberg and Goldstein, Your Old Country Tailors. Kibitzing Allowed." Unfortunately, neither Oakley nor Allan had time for kibitzing.

Oakley was draped in a white shirt, red bow tie, black tuxedo jacket with black silk trim, and black tuxedo pants. Several inches of the end of the pants legs lay on the floor.

The tailor, a man about fifty years old knelt in front of Oakley. A yellow cloth measuring tape and extraordinarily large faux gold crucifix hung around his neck. He held a thin bar of soap. Allan stood next to him. The tailor stood up.

"First, Stanley telling tailor joke," he said.

"Stanley, we're in a real rush," Allan pleaded.

"Stanley telling joke to all peoples. You want working?" Stanley had them cornered.

"Okay, okay, go ahead," said a resigned Oakley.

"So, tailor eating lunch and working too fast, spilling borscht on man's pants. Man comes to shop. Look at pants he saying, 'What doing this soup in my fly?'" Stanley chuckled at his joke.

"What?" responded a confused Allan.

"Uh, waiter is asked 'what's this fly doing in my soup?' Tailor is asked 'what's this soup doing in my fly?'" explained Oakley.

"Yah, that being right. Good joke, huh?" cried Stanley. "My joke."

"Yeah, good joke," sighed Allan.

Stanley realized, undoubtedly not for the first time, that he'd best keep comedy as a side career.

"Stanley being best tailor all Minsk. I now telling you needing cutting off two inches."

"How long?" asked Allan.

"I telling you two inches!"

"No, no. How long will it take to do the alterations?"

"What!? What being that?"

"Al-te-ra-tions," Allan fell into the trap of believing that by the speaker speaking slowly, a listener who does not speak English well could understand the meaning of unfamiliar words.

"How long to Stanley doing cutting?" Oakley asked, understanding that Stanley's English skills were limited.

"Oh, doing cutting. Sure. Stanley taking ten minute. That being all. Stanley being best tailor all Minsk. Ten minute."

"Good," said Allan.

Stanley looked at his wristwatch.

"Stanley being finishing two o'clock."

"Good. Great," exclaimed Oakley.

"Yeah. No joke. Stanley being finishing cutting being two o'clock. Tomorrow."

*   *   *

One thing about Salvation Army stores – they have large storefront windows. As with other clothing stores, those windows are meant to attract shoppers. If anyone enters a Salvation Army store, however, they do so not as casual shoppers but with this destination in mind. No one window shopping for clothing actually decides to enter a Salvation Army store. Vivian and Joan were the exception. After leaving the florist, they found the Salvation Army store around the corner.

Hundreds of dresses hung on racks from the front of the store to the rear. A saleswoman in her 60's held out a somewhat stained purple dress in front of Vivian as Joan watched. She gestured with her free hand and blabbered.

"This is very flattering. Tony Curtis wore it in 'Some Like It Hot.' Well, that's what Mrs. Pozinsky said. You know her son changed his name to Goldstein so no one would know he was Jewish. She's always bringing us things. But you never really know if she's just making stuff up. I remember a couple years ago she brought in an entire . . . "

"Lovely," Vivian said, trying to silence the saleswoman.

*   *   *

Pastor Elmer and Alice stood on the pulpit with Oakley, Vivian, Allan, and Joan. Pastor Elmer was now in official wedding chapel uniform, wearing a white and black clerical robe. Eleanor, in her 70s, sat at the organ, facing sheet music.

The trip to the tailor being partially successful, Oakley wore the white shirt, tuxedo, and bow tie. His pants legs were held up by safety pins. Stanley was very upset. Vivian wore a white dress and the ring from the key chain on her ring finger. They had mutually decided against hollowing out the pink gumball. Joan held a bouquet of flowers.

". . . and wife. You may . . .," concluded Pastor Elmer.

They had seen, no doubt, numerous spur-of-the-moment marriages, but Pastor Elmer, Alice, and the organist, Eleanor, had probably never seen the premature show of affection they were witnessing. Vivian and Oakley had been kissing in an extended embrace before being named man and wife in the name of whatever power Pastor Elmer invoked. It was perhaps understandable. The casino adventures, the near murder, and the rapid fire wedding preparation had electrified these nice young people.

Certainly, the pastor, Alice, and Eleanor had never seen the best man and maid of honor partake in the pulpit passion. Allan and Joan had joined Oakley and Vivian in their own extended embrace, as much out of passion as out of relief that the charade was over.

As soon as Pastor Elmer had finished pronouncing the marriage of Mr. Thorp and Mrs. Thorp, Eleanor had begun to sing "The Anniversary Song," accompanied by herself at the organ. She, too, had probably not witnessed such shenanigans and alternated looking back and forth at her music and the two couples. She flubbed some of the notes.

". . . kiss the bride," Pastor Elmer finished directing Oakley, in one of the great anti-climactic statements in modern nuptial history.

Pastor Elmer and Alice watched the two couples celebrate, right there, they must have thought, by their previously pure, white trellised pulpit. Alice breathed deeply. "If not for the grace of G-d," she must have thought, either longingly or thankfully.

Eleanor sang. Her thin voice was somewhat out of tune. Oakley and Vivian might have noticed if they had been paying attention.

Oh, how we danced, on the night we were wed,
We vowed our true love, though a word wasn't said.
Dear, as I held you close in my arms,
Angels were singing a hymn to your charms.

As she continued to alternate between looking at the couples and her sheet music, the distracted Eleanor made more and more errors in musicianship. One might have feared for her livelihood if indeed Pastor Elmer and Alice hadn't been similarly occupied.

Two hearts gently beating, murmuring low,
Darling, I love you so.

"Okay. That's good," said a breathless Pastor Elmer to the two couples, trying to regain control over the imminent desecration of his church.
"Okay. Okay."

* * *

"Okay," said Oakley. "Let's say I have a box with ten red balls and ten white balls. If I pull out a red ball I win and if I pull out a white ball I lose. What's the probability that I'll pull out a red ball and win? Fifty percent. Easy. And if I put that red ball back in, the probability I'll pull a red ball the next time is still 50%. But now let's say I don't put the red ball back in and so now the box has only nine red balls. Now the probability I'll pick out a red ball and win is only nine out of nineteen or about (pause) 47%. It's gone down because I didn't replace the red ball."

Oakley was in a good mood. Why not? He was now married to a woman he loved, and he had been a, relatively speaking, big winner at the tables. He was in lecture mode with an excruciatingly captive audience.

Oakley's red Ford Thunderbird was within the speed limit as they drove along Highway 50 on the first leg of their trip back to L.A. A box of unopened Uncle Ben's Converted Rice sat on the dashboard playing tag with his miniature dartboard and the red fuzzy dice souvenir. Oakley had suggested they might need the rice for dinner. Vivian sat next to him. They were both in their wedding outfits, Oakley in his tuxedo and Vivian in her white dress.

Vivian held a bag of oranges, a box of Oreo cookies, and some napkins in her lap. Allan and Joan sat in the back seat. Joan hugged the bouquet of flowers. Vivian was peeling an orange.

"Yep," Vivian said. "Very obvious." Hopefully Vivian was not only interested but attentive. If not, she faced an unhappy life with Oakley.

"Okay," continued Oakley. "Now take a deck of cards."

"Take my wife, please," said Allan.

"Not yet," said Oakley. "I still haven't made up my mind."

Vivian slapped him on the shoulder.

"I'm driving, dear," said Oakley, now in having-been-married-for-30-years mode.

"You can ask what's the probability I'll win a hand with a full deck of cards, like with the full box of red and white balls. Then, let's do this experiment. Remove a deuce. Now what's the probability I'll win a hand? Remove two deuces. What's the probability I'll win a hand? And so on. We've got to consider every possibility of cards being removed from the deck and see how it affects the probabilities of winning."

"Notice the smoke coming out of Edward's ears?" admired Vivian. "Here have a slice of orange."

She handed Oakley a slice of a peeled orange and a napkin.

"Thanks. I need the energy." Oakley hoped that Vivian remembered their words on the day following the Pacific Coast Collegiate Jazz Festival.

"You guys want some?" Vivian said to Allan and Joan in the back seat.

"Sure," said Allan.

"See, it's not like flipping a coin or spinning the roulette wheel where the probabilities always stay the same. The result of each flip or spin is independent of all those that preceded it and all those that will follow. Those are pure games of chance."

"Here," Vivian said as she handed two orange slices and two napkins to Allan.

"Thanks," replied Allan.

"But every time you remove a card, the probabilities change. The cards have a memory of their buddies who have left. The presence or absence of some types of cards have to favor the house, and the presence or absence of other types of cards have to favor the player," Oakley reasoned out loud, in furious thought. "Blackjack, for crying out loud, is not a game of chance!" Oakley ate his orange slice relishing both its freshness and his ability to devise a personification metaphor while both eating and driving.

"Hmmm. Great," Oakley judged.

"Gone to that big discard pile in the sky, so to speak," said Allan, appreciating Oakley's cleverness.

"Poor little cards," said Joan while caressing the bouquet and eating her orange slice.

"Yeah, so to speak," said Oakley, perhaps bothered that the Personification Monster he had created had evolved an existence of its own, distracting his students from the lecture.

"But that's the key to developing a more advanced system than that of Baldwin and his pals," Oakley persisted. "Their system is a very basic strategy. They don't take into account the changing nature of the deck."

"It's as if they're playing with an infinite number of decks of cards," opined Allan. He, of course, was fully aware of the significance of Oakley's words.

"That's exactly how you have to look at it. Baldwin treated the game as if it was sampling with replacement. But it's not. It's sampling without replacement, so the probabilities change, and we have to figure out which cards being removed from the deck favor the player and which being removed favor the house. Obviously, removing tens and Aces favor the house."

"You can't do all those calculations by hand," said Allan. "Well, maybe some guys have. At least Baldwin and his guys used a calculator."

"I'm sure some guys have done it by hand. Or tried to do it by hand. You'd have to deal yourself thousands of hands and keep track of what's going on. Maybe somebody did it analytically. Hopefully, I'll find some simplifications that will allow a shortcut. I think it's going to be fun."

Vivian tweaked Oakley's ear.

"I think being married to a mathematician is going to be fun."

"I wanna be married to a mathematician," said Joan, in a playful mood. "I wanna be married to a mathematician. I wanna be married to a mathematician."

"See what you've started?" Allan observed to Oakley.

8

The Ships that Go Sailing by Crickets

Nine out of ten Americans don't realize that the city of broad shoulders not only has a beach but also an extensive beach stretching twenty miles north and south along Lake Michigan. The master plan of city planner Daniel Burnham included yacht harbors, extensive parks, and a road parallel to the beach stretching from the far north side of Chicago to the far south side. Office and apartment buildings would not be allowed within nearly a mile from the Lake Front, as it was known. The drive along Lake Shore Drive appealed to bicyclists, walkers, runners, as well as those in automobiles. Pedestrian overpasses allowed beachgoers to avoid the traffic.

If you drove directly east into Chicago along Congress Parkway you would proceed under the massive post office building, into the Loop, past Michigan Avenue, and into Grant Park. Michigan Avenue compares to other grand boulevards of the worlds, with The Bund in Shanghai an appropriate reference. Both Michigan Avenue and The Bund are associated with waterfronts. If you stood in Grant Park and looked westward, you'd see a stretch of buildings from north to south. The Conrad Hilton, the Chicago Symphony building, and others were situated there, straight, in a row, forming a backdrop for the park rather than an intrusion.

At the end of Congress Parkway, where the road terminates with right- or left-hand turns onto Lake Shore Drive, was the answer of Chicago to the great fountains of Rome. Buckingham Fountain, designed by Burnham's collaborator Edward Bennett, is a nearly three-hundred-foot-diameter structure, with water-spouting stone classical figures directing their streams into the moat surrounding them. On summer nights, light shows of multiple colors emanating from underwater lamps accompanied the water jets, in well-designed

© Springer International Publishing AG 2017
L.M. Golden, *Never Split Tens!*, DOI 10.1007/978-3-319-63486-9_8

synchronism. Perhaps it was the phallic-ness of this display that attracted both tourists and lovers nightly, weather permitting.

The shoreline east of Buckingham Fountain was an embarkation point for tour boats. On summer nights such as this, with a gentle breeze creating gentle waves, deck lights from sailboats and yachts on Lake Michigan would flicker as the boats swayed in the water.

Julian and Nancy, holding hands, were among a crowd of about eighty people watching the light show. It ended with a display of towering streams of water, and the crowd applauded, although no light designer, hydraulic engineer, or choreographer was there to receive the accolades. Julian lifted their hands to his lips and kissed Nancy's hands.

It was only 10:30 or so, too early to abandon a nice evening, so Julian and Nancy went for a walk north through the rose gardens, part of the Versailles-like gardens of Grant Park also designed by Bennett, and then veered east to continue walking along the shore of Lake Michigan, all the while holding hands.

"Yeah, I know," observed Julian. "But it was over so fast. It was over as soon as it started. It reminds me of the brevity of life." Julian was finding melancholy in the evening.

"Oh, Julian," Nancy said.

"And the finality of death."

"It's Saturday night, Julian."

Julian persisted in wallowing in his mood. By this time in their relationship, he was comfortable sharing dark thoughts as well as mathematical frivolity, if there is such a thing, with Nancy.

"I remember reading *Of Human Bondage* by Somerset Maugham in high school. It tried to answer the question of 'what is the meaning of life?' Since then, really, I've thought about it a lot. I think I came up with a pretty good answer. I know it sounds contradictory, but I think the meaning of life is to gain immortality."

"Oh, that's good," Nancy half sighed and half spoke.

"We live so that we can find a way to never die."

"Oh, gosh. That's beautiful, Julian."

"Think about it. People get married so – well, one reason at least – so they can have children. Why do they want children?"

"To give their parents headaches." Nancy was willing to indulge Julian's reveries but a touch of humor never hurt. Julian chuckled.

"Right. Or to have someone take care of you in old age. But how about to gain immortality? Their genes, their likeness, lives on. Is there any joy for a person greater than seeing their kid with their freckles or their curly hair or their brains? That's immortality."

"*V'dor v'dor* in Hebrew. From generation to generation."

"That's why my mother loves you." Anne Braun in fact would tell her friends at Hadassah that Julian was dating a girl, a Jewish girl, who knew Hebrew and Yiddish. Julian kissed Nancy's hair.

"Some people give money to name a library after themselves," Julian continued, expounding on his thesis. "Some build up a stamp collection. Some write books. Some get prominent enough and lucky enough to have a screenplay written about them. These all achieve immortality, too. The vast majority don't have the money to fund a library or an interest in stamps or the brains to write a book, so they have kids."

"What did Somerset Maugham say?" Nancy had gotten entrapped by Julian's mind into what was apparently going to be the theme for the rest of their Saturday night date.

"He said the meaning of life is the pattern of life."

"Hmmm," Nancy hummed, not satisfied with the discovery of Mr. Maugham.

"Exactly. A pretty long book and not a very good answer. It's a tautology really. They mean about the same thing."

"Yeah. I love your mind, Julian." Julian had been able to use a term from logic to describe the conclusion of a great American novel.

Nancy pointed to a concrete ledge by the lakeshore. These concrete faux-boulders had to suffice for those wanting to view the lake from the Fullerton Avenue beach. It would have been too expensive for city fathers to have hauled in 10-ton boulders from glacial moraines.

"Here," she said, pointing to a spot.

They sat and kissed briefly. They stared out at the waves. Lake Michigan lacks the thundering surf of the east, west, and Gulf coasts, but on a night with a gentle breeze the surf would still break on the beach, spilling white foamy water over its crest. A bobbing sailboat passed by.

"I read that Baldwin paper," Nancy informed him.

"Very nice piece of work."

"Too many approximations," said Nancy, the mathematician.

"I agree."

"You can do better, Julian."

"We've just gotten a high-speed 7044 electronic computing machine at work," he declared. "The most advanced there is. I wouldn't have to make any approximations."

Nancy placed her head on Julian's shoulder. He raised her hand to his mouth and kissed it. The flicker of lights from bobbing sailboats accompanied the sound of crickets courting in the grass.

*   *   *

A loud chorus of crickets serenaded the night outside Oakley's rented coach house in Santa Monica. Insect mating seemed to be the rage this night from coast to coast, as the lyric from "Spring Can Really Hang You Up the Most" would say.

Street lights visible through the kitchen window illuminated trees against the night sky. At the kitchen table, Oakley sat with a calculator, a deck of cards, and a vase holding the wedding bouquet. The kitchen wall clock heralded 1:30 a.m., and, I guess, all's well.

The seven-hour drive along Highway 50 to Sacramento and then south on the interminable Highway 99 had taken nearly nine hours, Oakley, Vivian, Allan, and Joan taking frequent road stops to keep fresh. They had eaten dinner in Bakersfield, and it might have been chipped beef for the lack of culinary sophistication. Although Mr. and Mrs. Thorp had dropped off Allan and Joan and arrived at his coach house several hours ago, Mr. Thorp still hadn't gotten completely out of his tuxedo. He remained dressed in his tuxedo shirt and black tuxedo pants.

Oakley was writing on a sheet of paper, which served as a tally sheet. Vivian stood next to him, dressed in a nightgown and robe. With her arm around his shoulder, she leaned over the table and stared in disbelief at the tally sheet.

"You've done thirty decks?" she said.

"Just twenty more," replied Oakley.

"Being married to a mathematician sure is fun. Especially the honeymoon night," she lamented.

Oakley got the message, but still shuffled up the deck of cards.

"Well, maybe just ten more." He was in a creative trance that she would recognize many more times in their lives together.

"How's it look?" she asked. It was apparent that the honeymoon night activities would be temporarily suspended in the interests of science. She could accept his eccentricities. She had no choice, but didn't mind. It was part of the package of being with the man she truly loved.

"The result with the Aces removed was expected. I lost about two-and-a-half percent of my action. I mean, there's no chance for a blackjack and there's no splitting Aces."

Vivian twirled her finger in his ear.

"Just as expected."

"And we know removing tens from the deck will hurt the player, too. So I thought, what card is furthest removed from the one, that is, the Ace, and the ten? It's the five. So I removed all the fives." Oakley had at that moment discovered one of the basic theoretical concepts of what would become known as card counting. It would lead to the development of his first system, the "five-count strategy."

"Good. And?"

"Here are the numbers," he pointed to the tally sheet. Vivian stared at it, her eyes having troubling focusing, the fatigue of the day now interfering with her ability to appreciate Oakley's brilliance.

"I've dealt now just over fifteen decks with the fives removed and in all but four I won. And I've won significantly. Blackjack, darling, is not a game of chance."

"That's exciting," she said. "Very exciting." If she had to seduce him both verbally and physically, she would do so, as she began nibbling on his ear.

"You're following Baldwin's rules?"

"Just like at Barney's." He pointed to the tally sheet.

"Here."

Vivian blew softly in his ear and leaned over to look at the tally sheet.

"I'm ranging between winning two and six percent of my action," he declared with a mixture of triumph and amazement. "Average about three-and-a-half percent. Removing the fives gives an advantage to the player."

"That's fantastic," she agreed. "Exciting," she added, purposely using that term again. "You've got a lot of action here."

She put her arms around his neck.

"Well, maybe only five more decks," said Oakley, not immune to her seduction.

"Isn't it obvious what's happening?" she hinted.

Oakley turned around. "Yeah, I guess it is." He had surrendered to her siren's call, the dual meaning of "action" now being weighted more toward her figure rather than his figures.

"What's happening, Eddie my love?"

Oakley tenderly touched her cheek.

"I think being married to a mathematician is going to be fun."

9

Did You Hear the One About…

June, 1958

Try as much as they may, no amount of wax applied by the campus custodians to the floors of the Mathematics Sciences Buildings on the UCLA campus could hide its being a century-old building. The dark yellow painted walls remembered many students of the past, undergraduates, graduate students, and post-doctoral fellows. Many of those students may have wished that the building had a more Ivy League type name. Perhaps the walls did as well.

A piece of paper taped onto the closed door of Room 122 read:

10:00 a.m. Dissertation Defense
Edward Thorp
Compact Linear Operators in Normed Spaces.

Allan and Joan sat in chairs slightly down the hall from the lecture hall. Vivian paced.

"Besides striking out with the bases loaded and two out in the bottom of the last inning in the little league championship game with my twelve-year-old girlfriend Trish holding hands with another boy in the stands, this was the worst experience of my life," Allan recalled.

"Poor guy," responded Vivian. Joan had heard about Trish before.

"They humiliated me, but at least they didn't laugh at me. They look for a weak spot, then they throw a couple pints of blood in the water and it's shark-frenzy time."

Vivian placed the left side of her head onto the door.

"It's quiet in there," she noted.

© Springer International Publishing AG 2017
L.M. Golden, *Never Split Tens!*, DOI 10.1007/978-3-319-63486-9_9

"That's good," said Allan.

The blackboard inside room 122 was full of mathematical equations and geometric figures scribbled in chalk. Professor Taylor, Professor Robert Sorgenfrey, and Professor John Selfridge sat in chairs in the front row. Oakley stood at the blackboard with a piece of chalk in his right hand. Taylor was the head of Oakley's dissertation committee. The other two mathematics professors completed the committee.

"Bob, do you have any more questions for Mr. Thorp?" asked Professor Taylor to his colleague.

"I'm happy," replied Professor Sorgenfrey. Oakley was happy to have Professor Sorgenfrey on his committee, he being the individual who had referred Oakley to the Baldwin article.

"John?"

"Nope," answered Professor Selfridge.

"I've got one more question," Professor Taylor said to Oakley.

Oakley thought he had done pretty well. A dissertation defense is frequently a formality. The faculty advisor, in Oakley's case Angus Taylor, had guided Oakley's work throughout and was aware that it was suitable for awarding the Ph.D. Sometimes, rarely, a particularly difficult additional committee member would find a flaw in the proof, a lemma that required further substantiation, an assumption that was unwarranted, or some other problem. Such defects were usually not fatal. At worst, the candidate would be directed to provide additional steps in the proof.

Mathematicians are a rare breed. They are like classical pianists or chess masters. Either you have this ability to intuitively "see" the logic or you don't. Unlike other fields, such as astronomy, physics, biology, chemistry, botany, anthropology, sociology, political science, art history, architecture, and literature, where a large body of previous knowledge must be studied, in mathematics you need only complete a proof.

In the sciences, in particular, our minds are too feeble to deduce the countless manifestations of the laws of nature. It was the astronomer Sir Arthur Eddington, after all, who supposedly had said that if Earth were enshrouded in clouds we would never have known that stars exist. You must first observe the manifestations of nature, and then try to understand them in terms of those laws of nature. Acquiring Ph.D. level proficiency in most subjects requires, in short, the assimilation of a large body of knowledge.

In mathematics, in contrast, someone with the requisite intuitive logic skills can get a Ph.D. in mathematics in two or three years. State a proposition, see the solution, and do the proof. Anyone who takes ten years to get a Ph.D. in mathematics doesn't have the ability.

You have child prodigies in piano and chess for similar reasons. You either have the music in your head or have an advanced mode of both right and left brain processing that allows you to see dozens of moves ahead, respectively, in which cases you attain a professional level quickly, or you don't. One never hears of a child prodigy in astronomy or physics. One simply cannot master the body of knowledge that quickly.

Oakley thought he had done well enough to pass. Now Professor Taylor had another question. It could be a zinger. Maybe he had realized in the last two hours that Oakley had slipped up.

"Yes?" said Oakley, trying to hide his apprehension.

"Did you hear the one about the engineering, physics, and math graduates?"

Oakley laughed out loud. *Ah, a joke. Angus is going to tell a joke,* he thought. Oakley carried himself with dignity, just like Vivian. Most people, at first meeting him, would think he was almost intimidating. When, however, he laughed, he had a broad grin that would light up an entire room. It was inevitable that that gleam would lead to the end of this dissertation defense being full of levity.

"No."

"Shoot," said Professor Sorgenfrey to Angus.

"The college graduate with an engineering degree asks, 'How does it work?' The college graduate with a physics degree asks, 'Why does it work?' The college graduate with a mathematics degree asks, 'In which finite spaces does it work?' The college graduate with a philosophy degree asks, 'What do you mean by 'work'?' The college graduate with an art history degree asks, 'You want fries with that?' "

They all laughed uproariously.

The friends waiting outside room 122 heard the laughter.

"Oh, God!" said Vivian.

Allan buried his head in his hands.

"He's dead," agreed Joan. "He's dead."

"If they're going to give you a Ph. D, they want to make sure you're good enough so they can keep their reputations," lectured Allan disrespectfully.

"Unbelievable!" said Vivian, new to this nuance of academia.

"It's like being invited into a fraternity," said Allan.

"Oh, Edward," said Vivian.

"Without the partying after the initiation," said Allan, continuing the simile. "Well, we'll see about that party." Allan must have been reliving his unpleasant dissertation defense, and apparently imagining some kind of unrealistic retribution.

Inside the lecture hall, Oakley raised his hand, still in student mode. Some Ph.D.'s, having spent a decade or more as students, have recurrent dreams their entire lives about raising their hand to get the floor.

"I've got one," he said.

"Go ahead," said Professor Taylor.

"Okay. The surgeon and his team are operating on this guy. The surgeon says to the resident, 'Dr. Jones, how's the blood pressure?' 'Fine,' he says. He says to the anesthesiologist, 'Dr. Smith, how's the oxygen level?' 'Fine,' he says. He says to the intern, 'Dr. Reed, how's the spleen?' Dr. Reed says, 'Spleen!? Spleen!? I didn't know we had to know spleen!' "

Oakley's laugh was the loudest of the loud laughter of the four of them. He was now fully relieved that he had passed.

"Thorp, I hope you never have students like that," said Professor Selfridge. "By the way, do you know how can you tell the most recently hired assistant professor?" The others exchanged glances, this joke not being familiar. "He's the one without the knife in his back."

The raucous laughter continued.

Vivian didn't have to place her ear next to the door to hear the laughter. She pushed Allan aside.

"That's it," she exclaimed. "I don't care who they are or how badly he did. They're not going to mock him."

She strode to the door and grabbed for the knob. Before she could open it, the door swung open. Oakley walked out, with Professor Taylor's arm wrapped around his shoulder.

"Edward!" Vivian cried out.

"It's *dim sum* lunch for all," said a beaming Angus. "And it's on the department."

*   *   *

"That's right, Edward, if you had failed we would have all starved today," said Angus, making a presentation at Larry Toy's Restaurant in Westwood.

Ever since they sat down for lunch, the laughter had been non-stop. Chinese cuisine to these people didn't match the excitement of Korean or Vietnamese food, but Little Korea and Little Vietnam were miles east, near downtown Los Angeles, whereas Larry Toy's was one of the many Chinese restaurants catering to the student population in Westwood. It was also a tradition to have *dim sum* at the successful conclusion of a dissertation defense and that assemblage of small delights was only Cantonese.

If a Chinese student were being celebrated, he would have the chef prepare specialties known only to Chinese. That would be especially intriguing to the others, not knowing what kind of Oriental curiosities were being prepared. Still, the assorted dumplings, buns, rolls, rolled noodles, and steamed, boiled, and deep fried fish and meats fully massaged the receptive minds of the academics.

Tea cups, a pot of tea, and fortune cookies lay on a large circular table, the best way in which to serve the many *dim sum* dishes. Angus Taylor stood, delivering appropriate post-lunch niceties. Sorgenfrey, Selfridge, Oakley, Vivian, Allan, and Joan sat. Allan had brought a camera with an electronic flash attachment.

"So, that's why you guys churn out the Ph. D's," said Allan.

"The secret is out, Angus," lamented Sorgenfrey.

They all laughed.

"Okay," said Angus. "Now, to get serious. I have some good news and I have some bad news."

"You mean, I have only one hour of happiness?" said Vivian, relishing her introduction into mathematics collegiality.

"Well, okay," responded Angus. "Because Edward passed, let me first give the good news. Yesterday afternoon a telegram from MIT was delivered to Edward at his department address."

He pulled out a telegram from his pocket.

"I held it back until today."

"That was a wise decision," said Oakley. Knowing his application for a prestigious position had been received and acted upon would have no doubt distracted Oakley from his preparations for his defense, particularly if it had been rejected.

"Edward, I mean, Dr. Thorp, has been awarded a C. L. E. Moore Instructorship in the math department at MIT," said Angus. "It is one of the most prized positions in our field."

Everyone applauded and cheered. Allan nearly dislocated Oakley's wrist shaking his hand. Vivian gave him a hug. The professors each shook his hand.

"Oh, great," said Oakley. "Thank you, thank you. Great."

"You mean, my husband has a job?" asked Vivian.

They all laughed. She was fully accepted as a prospective faculty wife. It was time for more levity.

"You know the old joke about the pizza?" asked Selfridge.

"Go ahead," said Oakley.

"What's the difference between a sixteen-inch pizza and a mathematician?" Selfridge asked. Apparently the young people hadn't heard the joke.

Angus and Sorgenfrey had undoubtedly heard it, but they weren't going to spoil Selfridge's fun.

"The pizza can feed a family of four," delivered Selfridge. He almost stumbled on the punch line laughing at his own joke. He was definitely not Comedy Store material, that popular hangout for UCLA students down Westwood Blvd. from Larry Toy's.

Everyone laughed at the joke, the skill of the delivery notwithstanding.

"Seriously, Vivian," offered Sorgenfrey, "in academia we say a person has an appointment or position, not a job."

"That's because he doesn't get paid unless he also buses dishes at some campus fraternity, right Dr. Sorgenfrey?" noted Allan.

"That's not quite true, Allan," Angus corrected him. "He'll be . . ." Angus took a moment to read the telegram. He clearly had a better sense of comedy than his colleague.

" . . . busing dishes at the MIT faculty club," he concluded.

This was perhaps the best zinger of the afternoon, and they all laughed heartily.

"Much more prestigious," said Oakley in mock relief, to additional laughter.

Angus looked again at the telegram.

"Only lunches," he read. This guy was on a roll, as the continued laughter testified.

"Good," noted Vivian. "Then you'll be home for dinners."

"You said there was bad news," said Oakley, reminding Angus of his introductory comment.

"Ah, yes," recalled Angus. "I'm sorry to say, I'm very sorry, Edward, to have to say, that . . . " he hesitated.

Oakley and Vivian were understandably quiet. They had to be wondering if this was another set-up to a joke or did Oakley's advisor in fact have some crushing information affecting Oakley's future.

Angus continued, ". . . as a result of obtaining this position Dr. Thorp will not be staying on at UCLA beyond his one-year teaching appointment as a research post-doc."

"That's bad news?" mocked Allan.

This time Oakley led the raucous laughter. He almost felt like strangling the good professor.

"I would like to propose a toast," said Selfridge, standing. It would have been appropriate for Angus to propose the toast, but he was still laughing at the success of his delivery. He gestured for the rest to stand.

"Hear, hear!" shouted Angus, rising.

Sorgenfrey, Vivian, Allan, and Joan stood up with their tea cups raised. Only a group of academics would make a toast using tea cups, Oakley must have thought. He stood up to join his friends, not really understanding the protocol.

"Angus?" said Selfridge knowing that Professor Taylor would be, as I said, the appropriate one to make the toast.

"Thank you, John," said Angus. "Dr. Thorp, you have been a brilliant student," he said. "We know you will have a brilliant career. To the best of health, happiness, and prosperity to you and your lovely wife in your new position and beyond. Cheers!"

Together they all joined in a chorus of "Cheers!"

Vivian hugged Oakley, the group offered another round of handshakes, and Vivian hugged Oakley. Seeing this, Allan quickly grabbed his camera, stood back, and took a photograph of the two of them.

"Anybody got a taste for a sixteen-inch pizza?" he asked.

10

A Coming Out Party

January, 1961
Willard Hotel
Washington, D.C.

Academic annual meetings differ from their counterparts in the other professions and the business world. Medical association meetings, for example, serve to announce advancements in equipment and therapies, provide a job fair, and create a forum for vendors to promote new techniques, software, and diagnostic equipment. Corporate meetings serve to instruct sales staff in new products and selling techniques, introduce new promotional material, and stimulate morale, money being the bottom line. Entertainment of some sort is provided, I believe, to stimulate the inevitably waning enthusiasm of the sales staff. Non-corporate business meetings, from my admittedly meager experience, are basically vacations.

Academic meetings have other functions, all associated with colleges and universities and research institutions. First, they provide undergraduates an opportunity to make their first presentations. Second, they provide junior faculty an opportunity to expand their resume, hopefully leading to tenure at their present or some future school. Vendors will be present, including book publishers and developers of whatever educational and research aids are appropriate to the given discipline. Job fairs are a major attraction, both for the recent graduates and hiring institutions.

The overall goal is to disseminate the latest research. Academia is specialized, with faculty and researchers in one area having limited knowledge of the latest results in other areas. A typical session will have, depending on the topic,

© Springer International Publishing AG 2017
L.M. Golden, *Never Split Tens!*, DOI 10.1007/978-3-319-63486-9_10

a couple dozen to maybe fifty attendees of those doing research in that area. Although it's challenging enough to keep up in your own specialty, these meetings provide academics the opportunity to learn about advances in other specialties. From a social standpoint, meetings provide a common venue for former colleagues, either as students or faculty who have gone to other institutions, to reconnect. These meetings, as a result of these reconnections as well as the finely tuned senses of humor, sparkle with joviality.

The 1961 forty-fourth annual meeting of the American Mathematical Society in Washington, D.C., was typical of such gatherings, with one exception. A young professor from MIT was presenting a paper on a strategy for the casino game of blackjack. Anticipation of this event had led to articles on January 25, the day before his Thursday, January 26, presentation, in the *Washington Post*, *Boston Globe*, and other papers, the content of which was quickly picked up via news services by thousands of other newspapers across the country.

After his encounter with Baldwin's system during his trip to South Lake Tahoe with Vivian, Allan, and Joan, Oakley had spent more than one hundred hours playing mock hands of blackjack, studying the effect of removing one or more of the Aces, then removing one or more of the twos, through all ten types of cards, the tens, Jacks, Queens, and Kings being all considered ten-value cards. It had taken Baldwin, Cantey, Maisel, and McDermott thousands of hours over one and one-half years, working evenings and weekends, to develop just their basic strategy. At Oakley's request, Baldwin had sent him the notebooks of his group's calculations, enough to fill a cardboard carton. As he studied them over weeks in the spring of 1959, Oakley became further convinced that the Four Horsemen hadn't discovered the hidden secrets of the game. Their strategy, while sound, only considered the dealer's up-card and the player's two cards. He tried, working in a trancelike state, but he eventually admitted to himself that he had taken on a problem too complex to solve in his lifetime, even in a simplified form. He gave up, a crushing decision for someone as competitive as Oakley.

After beginning his duties in the summer of 1959 in the C. L. E. Moore position at MIT, however, he was given permission to use their IBM high-speed 704 electronic digital computer at the MIT Computation Center. Now he had a real electronic computing machine with which to perform his calculations. He was sure he'd have to make some approximations, but thought this venture would provide great scholarly rewards and great fun.

First, he had to learn how to communicate with the 704 electronic digital computer. As many bright professors and graduate students in the sciences would do beginning in the 1960s, Oakley taught himself the FORTRAN programming language, short for "formula translation." The program was designed for mathematical analyses.

Using this great tool, much more sophisticated than the Friden electrical-mechanical calculating machines that Baldwin and his group had available, he joyfully and systematically programmed the effect of removing cards of various values from the deck. It became clear, without laboriously dealing mock hands for hundreds of hours, that removal of the low value cards, the twos through the sixes, provided an advantage to the player, and that removal of the ten-value cards and Aces provided an advantage to the house. Such a possibility had hit him after trying Baldwin's system on that trip to South Lake Tahoe with Vivian, Allan, and Joan, reinforced, briefly, on his honeymoon night, and tantalizingly indicated in the more than the mentioned one hundred hours of dealing himself mock blackjack hands over eighteen months while he remained in L.A. after the Tahoe trip.

He had proceeded systematically. First, he wanted to use the 704 electronic computer to discover the improvements he could make over Baldwin and his group's system. It was obvious that his use of a much more powerful computing tool than their hand calculators would provide some if not many. He began with a full 52-card deck.

The necessary initial data were the probabilities of the dealer getting a given total for a given up-card, knowing that the dealer has to stand with a total of 17. If, for example, the dealer had an up-card of 5, Oakley had to determine the probability that after drawing the dealer would end up with a total of 17, 18, 19, 20, 21, or bust. To do this by hand to obtain results with reasonable precision would require dealing thousands of hands. That's the chore that Baldwin's group faced. With FORTRAN and the 704, it required just twenty or so lines of code, and he could have the computer deal as many hands as he wanted until no significantly different result would be obtained by dealing more.

The results were startling. If the dealer's up-card were a 4, 5, or 6, Oakley found that the dealer had a 40%, 43%, and 40% chance of busting, respectively, nearly half the time! If the up-card were a ten-value card or Ace, on the other hand, the chances were only 21% or 12% of busting, but the dealer in those cases had a 37% and 52% chance of obtaining a total of 21. With only this knowledge, the player would be encouraged to split and double down more frequently against a dealer showing an up-card of 4, 5, or 6.

These results provided the foundation for all the subsequent calculations. To determine, for example, the player's advantage for standing on a given total, one would simply add up the probabilities that the dealer would fall short of that total or bust and subtract the sum of the probabilities that the dealer would exceed that total.

Thus, to determine the player's advantage for standing with an 18 against the dealer's up-card of 5, Oakley would add the probabilities that the dealer would obtain a 17 or bust, which he found to be 0.120 and 0.429, respectively, for a sum of 0.549, and subtract the probabilities that he would obtain a 19, 20, or 21, which he found to be 0.117, 0.105, and 0.106, respectively, for a sum of 0.328. The total would be 0.549 − 0.328 = 0.221, indicating that the player had a 2.21% advantage by standing with an 18 against the dealer's up-card of 5.

In general the player also has the option of taking a card to improve his or her hand. Oakley therefore also programmed the 704 to determine the advantage the player would obtain by drawing cards. This would be a little more complicated. In our example, the player could draw an Ace, 2, or 3 to improve the hand, whereas drawing a card with a higher value would cause him or her to bust. The same summing of respective probabilities would yield the advantage.

If, for example, the player were given a 3, giving him a total of 21, then the probability that he would win against the dealer's up-card of 5 was the sum of the probabilities that the dealer would get a 17, 18, 19, 20, or bust times the probability that the player would actually get that 3, minus the probability that the player would bust, getting a card with a value greater than 3. The same calculation would have to be performed if he had drawn an Ace or a 2, and Oakley would then add the results of the three calculations to determine the advantage of drawing the card.

The difference between the advantages of drawing or standing would determine the player's strategy when faced, in this example, with an 18 against the dealer's up-card of five. If the player didn't bust, he had a further option of drawing another card or standing.

He similarly programmed the 704 to determine the advantages to the player of doubling down with given totals against the ten possible up-cards of the dealer, and the advantages to the player of splitting pairs, again against the ten possible up-cards of the dealer. Taken together, for every dealer up-card, the player with any two cards could now compare the advantages for splitting, doubling down, and drawing or standing to see which was the most advantageous strategy.

No wonder it took Baldwin and his gang thousands of hours to do this by hand calculator.

Now Oakley faced a creative conundrum. The drawing and standing table was a matrix of ten possible dealer up-cards and nine possible player's hands, 12 through 20, a player having less than 12 always taking another card. That was a total of ninety entries. The hard and soft doubling down tables and the splitting table added dozens more. Because he couldn't expect anyone to memorize 150 to 200 decisions, he had to determine a means to organize the results into something more congenial to actual casino play.

He followed Baldwin's lead. First, most of the decisions are obvious. You don't double down on a sixteen. You don't split fives. You don't take a hit on a twenty.

He further simplified the results for actual casino use by combining all decisions for a given dealer up-card. The logic was simple. If the results showed that a player should hold with a 17 against the dealer's up card of ten, then the results showed the obvious, that he should hold with an 18, 19, 20, or 21. If the results showed that the player should double down with a ten against the dealer's up-card of nine, then he should also double down against the dealer's up-card of two, three, four, five, six, seven, and eight.

With these simplifications, the result could be summarized in concise tables that the player could memorize with what Oakley considered an acceptable, if not minimum, effort. Oakley, with the quick mind of the 704 as his partner, was revolutionizing the game of blackjack.

Taken together, Oakley's more precise results showed that the player in a typical casino had a 0.12% advantage over the house. If he followed the directions that Oakley summarized in the concise tables, for every $1000 wagered the player would in the long run make $1.20. Oakley referred to the refined system as the "basic strategy." The results showed that Baldwin's group, while groundbreaking pioneers, had erred in their hand calculations slightly, having concluded that it was the house that had the small advantage.

They had published a result that the house would have a 0.62% advantage, which they later found, as communicated by Wilbert Cantey to Oakley, was incorrect as a result of an arithmetic error and that the house advantage was actually only 0.32%. Oakley redid his computer calculation using the strategy they had published, and in fact found that their system provided an advantage to the player of 0.09%, remarkably close, attesting to the magnitude of their efforts, to Oakley's electronic computer-generated results.

Because the system as published by Baldwin and his colleagues, and corrected, gave the house that small 0.32% advantage, a nearly even-money game much better for the player than any other casino game, the implication was clear that a skillful and somewhat lucky player could win at blackjack. Oakley had now shown that in the long run the player did not need skill or luck, but would indeed win, although at a miniscule rate.

All that work to the good, Oakley's results were basically a validation of Baldwin's results, and Oakley had yet to incorporate blackjack not being, as he had noted to his friends after the marriage ceremony, a game of chance, that the trials in a game of sampling without replacement are not independent. He had to calculate using the 704 the effect of removing various cards.

He redid the analysis for removing one, two, three, and four of the fives. His results were astounding. The results showed that removing the fives provided up to a 3.6% advantage for the player. To any casual observer, using a favorite expression of scientists, this was a major, major advance. Wager $1000 and you expect to win in the long run $36. He referred to the strategy based on tracking the number of fives dealt from the deck as the "five-count strategy."

Astonished, invigorated, ebullient, he carried on relentlessly. He began to study the effect of removing ten-value cards from the 52-card deck. He would proceed just as he had with removing the fives, but now he would have to calculate the effect of removing 16 cards. As he had noted to Vivian in her nightgown, an abundance of ten-value cards and Aces compared to a neutral deck would increase the chances of blackjack. Because the dealer has to take a hit on 17, their presence increases his or her chances of busting. The player, who is not so constrained, can decline to take a hit holding cards which total, for example, 16, knowing that the deck is rich in ten-value cards. The presence of those cards, Oakley thought, should also direct the player to double down and split more frequently.

On the other hand, the presence of low-value cards would help the house. The dealer has to take a hit below seventeen. He or she would have a better chance of drawing one or more low-value cards, getting a 20 or 21. Oakley reasoned that when the deck is rich in low-value cards compared to a neutral deck, the player should double down and split less frequently, and take a card rather than stand with poor hands.

It was becoming increasingly clear to Oakley that having the deck relatively rich in low-value cards decreases the player's advantage, but having the deck relatively rich in high-value cards would increase his or her advantage. He had, accordingly, begun programming the 704 for a more advanced system based on tracking the ten-value cards that had been dealt. That would be called the "ten-count strategy." He estimated that, assuming it takes 10 seconds to play a hand in the casino, the 704 calculations, which played more than 30 billion hands, would have taken 10,000 man-years of actual casino play to accomplish.

It was late 1960, however, and he wanted to share the results of his investigations based on removing the fives with colleagues. The completion of this analysis would have to wait.

He had submitted a paper on the "five-count strategy" system to the American Mathematical Society annual meeting in Washington, D. C. The paper had been accepted for presentation. Oakley was only mildly surprised at this. After all, he was on staff at MIT, and the content of the paper was obviously not quackery, despite its dealing with a casino gambling game.

An Associated Press editor in D. C. perusing the program for the meeting saw the beguiling if somewhat presumptuous title of Oakley's presentation, "Fortune's Formula: The Game of Blackjack," and directed AP staff in Boston to write an article about this young mathematician who promised gold that could be mined in every casino using only mental picks and shovels. A *Boston Globe* writer by the name of Dick Stewart called Oakley in the late evening of January 24 for a phone interview. A photographer was sent to his home.

As I said, the next morning, Wednesday, January 25, a front page article appeared in the *Globe* and quickly thereafter newspapers across the country. The page 3 *Washington Post* article by Thomas Wolfe the same day, based on a second telephone interview with Oakley, quoted Oakley as saying, "Frankly, I would welcome a wealthy backer or a gambling casino itself to finance me in what we might call the acid test of my theory. I haven't been able to accumulate the capital to undertake it on my own."

Almost instantaneously, the phone calls from rabid gamblers began. They continued throughout the day and night from all over the country. By 4:00 a.m. Oakley was a wreck. He had found himself in a world as foreign as those of other planets of which he, as an astronomy buff, had wondered, but he knew he wanted to test the system and was going to present his paper and see the thing through.

The venue in the Willard Hotel where Oakley was to give his presentation was typical of academic meetings, a large ballroom divided into spaces for three separate academic presentations by portable partitions. It was in fact the appropriately named Grand Ballroom. On the slightly raised stage that served as a dais, a sign on the front of the lecture podium for the speaker provided the name of the meeting and the hotel,

Willard Hotel
American Mathematical Society, 1961
Washington, D. C.

A large screen behind the speaker displayed slides, and the speaker held a push button device to allow him to guide the slide projector to display the next or previous slides.

What distinguished Oakley's presentation from all others at this meeting was the number of those attending. More than one hundred people filled the chairs, others stood along the walls, and still others stood in the hall outside the room. This was, I repeat, a mathematics convention. The press coverage had generated major interest, and was to change the direction and import of Oakley's life.

Oakley stood behind the podium. The screen behind him displayed a slide showing tables of numbers and a large multi-colored pie chart. Each table had the title "The Five-Count Strategy for Blackjack."

The moderator for the session stood to his side. Incredibly rare for an academic meeting, two photographers were taking photographs of Oakley, their flashbulbs popping. Julian Braun sat in the first row, wearing a name badge identifying him as a registered meeting participant.

"The question was, although the five-count strategy works, doesn't the low rate of winning make it impractical," said Oakley, directing his comments to the audience. "It's certainly true that the rate of winning is low. You have a three-and-one-half percent advantage over the house if all four five-value cards have been played." He pointed to the slide on the screen. "But as my table four shows, that only happens about six times out of one hundred hands that you play."

The session moderator pointed to a man in the audience.

"Question for Dr. Thorp."

I always wondered at the use of the term "moderator" to refer to the facilitators of sessions at an academic meeting. It reminded me of the moderator rods inserted into nuclear reactors to absorb neutrons and control, that is, moderate, the fission rate and therefore power output, preventing annihilation of life on Earth. At meetings, the "moderator" would presumably prevent uncontrolled disagreement between academics leading to intellectual rage and consequent cerebral meltdown.

The man in the audience spoke. "It's easy to keep track of the number of fives that have been played," he said. "Wouldn't appreciably more complicated systems be impossible to memorize? Would you have to essentially carry around a portable computer?"

Members of the audience laughed at the absurdity of such a concept. In the early 1960s the only computing machines were main frame computers which, because of the primitive electronics, literally filled entire rooms.

"That's a good point," admitted Oakley. "The question was essentially that if systems more sophisticated than the five count strategy get developed will the player be able to manage the calculations in his head or will a portable computing machine be necessary?"

The person asking the question nodded in agreement.

"I think both will occur," continued Oakley. "I think the most complex systems that will be developed will require a portable computer. However, simplified versions of the systems would be amenable to mental calculation. In fact I'm now working on a system which I call the ten-count strategy. Here the player tracks all the ten-value cards that have been played and computes the ratio of non-ten-value cards to ten-value cards remaining in the deck. It's more complicated than the five-count strategy, but it can still be memorized."

The moderator pointed to another man in the audience.

"Question for Dr. Thorp?" he asked.

"Sure," said the man in the audience. "If you're a rocket scientist."

The audience laughed at this remark, not because it was inappropriate but because the man, in his non-academic manner of speaking, was thought by most in attendance to probably be correct.

Although those presenting papers or sponsoring booths typically must pay a fee to attend academic meetings, the publicity generated for Oakley's presentation had persuaded the session organizers to allow the general public to attend.

"You admit there's a low rate of winning, and maybe you'd need to carry around a machine on your back," the man continued. "Tell me, these electronic machine calculations are nice, but I don't think it'll work in practice. Have you left your ivory tower in the sky to really try any of your so-called systems in a real casino, Mr. Thorp? I mean, a real casino."

Members of the audience laughed and Oakley cringed. He had fought to keep his session open only to the press and academics, but the session organizers prevailed. Oakley was now paying, having to endure the unenlightened commentary from one who was obviously not in the field of mathematics.

"No," said Oakley.

The audience laughed again, this time with a bit of mockery. Oakley was getting upset. He felt that the moderator should have put this fellow's questioning at rest.

"Look, it's mathematics," Oakley said, now feeling he was back in the neighborhood facing an imminent street fight. "The mathematical details will be in my paper. It works. I don't have to prove anything to you. It works."

The audience responded to this with increasing laughter. A few catcalls were heard. The moderator decided it was time to terminate the presentation.

"One more question for Dr. Thorp," he said.

Julian had been listening to the interchange and had raised his hand. The moderator, noting that Julian wore a registration badge and was most likely not a layperson, pointed to him.

"Can you estimate the errors in your expectation values resulting from the approximations that you made in your computer program?" he asked.

Oakley was glad to finally get an intelligent, relevant question. Hopefully the non-mathematician badgering was over.

"Yeah," he said. "Actually, no. I would hope that the errors are small if they exist at all. My programs are free from bugs, I know that. And the system works. I can't estimate the errors because more powerful systems haven't been developed yet." Oakley was still irritated, and he was unhappy that he wasn't able to properly answer the question.

"Why don't you do it?" he challenged Julian.

The audience felt that the prey had been cornered and was fighting for his life. They laughed and uttered more catcalls.

"With that," the moderator said, "We're going to take a twenty-minute break before our next session."

The audience rose from their seats, and various groups began to converse. Oakley removed his slides from the slide projector, gathered his notes, and walked off the dais.

A group of about fifteen to twenty descended on Oakley. Oakley looked at them hoping they weren't vultures and hyenas coming to remove what was left on the bones. He had hoped the attacks were over. He feared, however, that they would begin anew. Some of them, judging from their ready notepads and pencils, were reporters. Julian stood to the side.

"Dr. Thorp," said one of the reporters. At least he's addressing me with respect, thought Oakley. "Joe Sherman, *The New York Times*. You said that blackjack players could win regularly with your system. Can you translate that into cash?" It was a reasonable question.

"It depends on your minimum wager," replied Oakley. "If it's one dollar, I think winning one hundred dollars in an hour playing heads-up against the dealer would be possible."

"Dick Stewart, *Boston Globe*," said another. Oakley remembered the name as the reporter who had written a long piece about him and his work in the *Globe* that morning.

"That's a very optimistic claim," said Stewart. "I'm going to ask the same question that was asked before. You haven't actually tried your system in a casino. Do you ever plan to?"

"Oh, hello," said Oakley, trying to stay on good terms, despite the potential dangerous implication of the question. "My biographer. Thanks for the article in today's paper. At least that was nice. To answer your question, Dick, I haven't tried the five-count strategy yet. I've tried the system developed by Baldwin and his colleagues and the results were very suggestive."

"But you haven't tried your system," persisted Stewart. "And what do you mean by 'suggestive?'" Oakley, to his dismay, had apparently found a new enemy. He tried humor.

"I tried it in Lake Tahoe when I was a graduate student and didn't lose all of my money," joked Oakley, smiling.

At least the group laughed, and Stewart was silenced.

"Actually, I made $8.50," Oakley continued. "But everyone else at the table lost all of theirs."

"Allen Weinrub. *Washington Post*," said another reporter.

"Dr. Thorp, you haven't tried your first system in a casino. Do you plan on soon trying out your new system in a casino? The ten-count strategy." That was a reasonable question.

"I don't think so. I'm a mathematician, not a gambler," Oakley replied. He had discovered an effective way of deflecting the embarrassing questions about trying out the system in the casinos. "It's not my world. And remember, in the short run you can lose money. Even if I had the urge, I'd need to be bankrolled. University professors can't afford to lose $500 in a casino."

"Our editorial today read," continued Mr. Weinrub. He started to read, and it was apparent he was going in for the kill.

"There's a mathematician in town who says he has a system for beating blackjack. It reminds us of an ad we answered. Send only one dollar for a surefire weed killer. Back came an envelope with a piece of paper saying 'Grab by the roots and pull like hell.' Would you like to comment on that?"

Oakley tried the humor-deflection technique.

"I'm mathematician, not a botanist."

The group laughed at this witticism. His theory may not work, they might have thought, but he's a good enough guy.

The first reporter, Sherman, wouldn't relent.

"How do your colleagues at MIT view these pursuits of yours?" he asked. "There's a lot of skepticism in this hall about the whole thing. How do they feel?"

"Well, they're not willing to bankroll me, if that's what you mean."

Some of the group laughed, but they were getting tired of Oakley acting like a sleaze politician, avoiding the questions. Some muttered non-niceties under their breath.

"The students love it," said Oakley sensing the growing antagonism.

Stewart dove in. "Is it true, Professor Thorp," he said, with notable sarcasm in referring to Oakley by his professional title, "that your colleagues in the math department view your non-traditional research unfavorably and that far from getting tenure your contract for next year may not be . . ."

Oakley had endured enough. The humor didn't work. The sympathy-engendering comment about low faculty salaries didn't work. The reference to his admiring students didn't work.

"No more questions," he said, walking away. "Thank you very much."

"Dr. Thorp . . .," said Sherman, wanting more.

"No more questions," said Oakley.

The three reporters thanked Oakley and let him go as the group dispersed. Oakley walked quickly away through the ballroom/lecture hall. Julian ran up to him and confronted him in the aisle.

"I think you're wrong," said Julian. "The approximations you had to make, even if incorrect by only two or five percent, would dominate your expectation values."

"Excuse me?" replied a nearly shellshocked Oakley.

Oakley continued to walk and reached the hallway outside the ballroom. Julian followed him like a new puppy.

"I read the Baldwin paper, too," he said. "I think your counting fives is a major advance over their results."

"Why, thank you," said Oakley, his despair over his treatment in the hall transforming into sarcasm.

"But you must have used a 704," said Julian. "It's obvious from all the approximations you had to make. You're hampered by it like the Baldwin group was hampered by using only their Friden calculators. And your code is probably inefficient."

Julian, not the most socially adept, was perceptive and obviously knowledgeable. The quality of Oakley's work was in fact diminished by his having to use the computer available at MIT, and he had needed to make approximations.

"Then, my friend, like I said, develop a new system . . .," said Oakley looking at Julian's name badge, "Mr. Julian Braun." Oakley should have been more receptive to Julian, although his rudeness was understandable after his treatment in the hall. He knew that Julian had accurately analyzed the quality of his work.

"Mr. Julian Braun, M.S," said Braun, standing up to Oakley. "You're not the only intelligent person in this hall. I work at IBM's facility in Chicago. I have access to a 7044." That was Julian's final attempt. He had shot the biggest shell in his arsenal right at Oakley.

Oakley stopped walking. He had been struck.

"A 7044?" he said with a glint in his eyes.

"I'm not interested in gambling. If you let me take a look at your code, I'm sure I can avoid the approximations and shortcuts," added Julian, firing another computer salvo.

"A 7044?" Oakley was thinking of the consequences of using such a machine for his work.

"That's right. I'm a programmer. I'd like a copy of your FORTRAN code."

"A 7044?"

11

Braun's Brawn

The office of the MIT math department on the first floor of Building 2 was in room 2-106. This in itself would pose an enigma to the first-time visitor. Clearly, a room with the name 2-106 would be on the second floor. When Oakley first came to the campus, he expected the mathematics building to have the name of a wealthy New Englander who had endowed the university with a large gift. Instead, he found the name, literally Building 2, to be even less enchanting, if possible, than the name Mathematics Sciences building that he had known as a student and then in his teaching position at UCLA.

The office, 2-106, which was on the first floor, was decorated in neo-classical musky, with brown painted plaster walls with some narrow wood trim and several metal desks. A clock on the wall read 9:15 a.m. The calendar said it was Monday, as if those coming to the math department after the cold Massachusetts January weekend didn't know.

As all academic offices are, this office was functional. A swinging door would open at the far right end of a counter stretching from left to right a few feet beyond the door to the hallway. At the left end of the counter a large rectangular metal box attached to the wall housed a mailbox. It had twenty-four mail slots with pieces of paper with the names of faculty members glued to the base of the respective slots. Underneath those slots were about forty smaller slots for the graduate students and post-doctoral fellows.

Magazines, letters, and note slips sat in the various mail slots. Oakley's mail slot was empty. Oakley stood at the mailbox. He stuck his hand into his mail slot and moved it around to make sure no piece of mail was lying flat in it. Jane Rose, the department secretary, sat at a desk on the other side of the

© Springer International Publishing AG 2017
L.M. Golden, *Never Split Tens!*, DOI 10.1007/978-3-319-63486-9_11

counter, typing. She was perhaps 5′2″, 30 years old, with a ready, deep, nearly baritone laugh and trim figure. Everyone liked her.

"No mail?" asked Oakley.

"Oh, yeah," Jane replied.

"My mail box is empty," noted a puzzled Oakley.

"Oh, it's in the sack behind the counter."

"Sack?"

"There was way too much for your mailbox."

"Anything important?" Oakley was surprised by Jane's comment, but then realized his talk and the newspaper coverage may have generated some responses.

"No," said Jane. "Just the usual invitations to appear on What's My Line, To Tell the Truth, and Truth or Consequences."

"Really?"

"And a couple hundred reprint requests for your upcoming paper on blackjack. Maybe a thousand. Who counts?" You could see Jane's personality shining. She continued. "Letters from gangster and bookie types. To take a class on gambling no doubt. Fan mail. Some are female admirers judging from the clouds of perfume."

"Stratus or cumulonimbus?" said Oakley, getting into the spirit of the morning. The unaccustomed stress of the days leading up to his presentation, and the grilling after his presentation, had tired his body and dulled his spirit, but after a weekend back in Brookline with Vivian he was refreshed, and his smart-aleck nature was rejuvenated.

"Oh, you lady-killer mathematicians."

"And there's a man who called and says he's an electrical engineer and says he developed a better system than your system."

"Oh, how nice. What's his name?"

"Harvey Dubner. He says he's going to send you a copy of his work."

"Okay. Anything else?"

"Oh, this is the really important one. Vivian called and said don't forget the bananas and milk."

"Thank God. I almost forgot. Jane, can you take care of the reprint requests?"

"Already done."

"What a dear," said Oakley, expressing his affection for Jane. "The astronomy department's loss is our gain."

"Well, we've got celebrity here."

She handed him a handful of pink message slips and a small stack of letters.

"Here are the phone call slips and what I think are the mobster letters," Jane said. "I was almost, almost I say, tempted to open them up. I'm sure they make great reading."

Oakley began to walk out the door with the letters.

"They might make excellent students," he said. "But you probably made a good decision. Your life may have been at risk." He stopped walking. He handed Jane some of the letters.

"Here, you can keep the perfumed letters. My life would be at risk if I took them home."

"Sure," agreed Jane. "There were some phone calls that, well, weren't very nice. Critical, you know." Oakley could have expected that.

"And there's one guy who keeps on calling. He doesn't sound like a gangster. He says he wants to visit you next week."

"Julian Braun from Chicago?" Jane was taken aback.

"You can read minds? Mathematicians are taught how to read minds?"

"Only some of us. It was an honors course."

*   *   *

The blackboard in Lecture Hall 112, Building 2 – officially room 2-112 – was full of drawings sketched with chalk of square grids similar to rectangular graph paper. One drawing had a conically shaped figure drawn to simulate it penetrating through the graph paper, with the words "wormhole" identifying the cone. Next to it was a drawing with the graph paper drawn as if it were bending around on itself, with the wormhole cone shrinking and then enlarging and opening back onto a different part of the drawing. Numerous equations were written next to the drawings.

A clock hung on the wall above the center of the blackboard. It was five minutes to eleven o'clock. Oakley's lecture notes lay on a non-descript wooden lectern in front of the blackboard, and his briefcase was propped up next to the lectern. Oakley stood at the blackboard holding a piece of chalk, addressing his twenty-five or so students sitting in the classroom. Julian, cradling a bulky, large manila envelope in his lap, sat in the back row.

"The two spaces are equivalent mathematically, simple topological transformations," said Oakley. "Now, physically, they represent different phenomena, as you can see." He pointed to the first diagram. "Here the singularity is isolated. Here," he said pointing to the second diagram, "its nodes are located in two different regions of the manifold. Separated by arbitrary distances depending on the value of the stretching parameter, kappa. But it's the same singularity."

He looked up at the clock above the blackboard.

"Oh, I'm running over. We'll, we'll talk about the possibility of using those wormholes for interstellar travel next time. I assume you are interested in interstellar travel?" he teased.

The students rose from their seats and left. Oakley placed his lecture notes into his briefcase as Julian walked up to him.

"Good stuff," Julian observed.

"It keeps them from falling asleep," said Oakley wryly. He pointed to one of the students leaving the lecture hall.

"But that Alan Guth kid. Bright kid."

They walked out of the lecture room, into the corridor, and down the hall into Oakley's office.

"It's good that the physics students take these classes," he continued. "Shows the math students that there's a reason for their existence."

Julian laughed. "Right. Math is the language of physics."

"I just wish some of the finance or economics students here would enroll," stated Oakley. "These are powerful tools. I'm sure they could be applied."

"Why don't they enroll?" asked Julian.

"I think the finance majors are too busy picking stocks," claimed Oakley, "and I think the economics students are too busy working with their faculty advisors, falling over each other with their hare-brained theories trying to win Nobel Prizes."

"There's a great word in Yiddish for that," offered Julian. "*Fakakta* theory."

"*Fakakta*," Oakley let the word staccato out of his mouth. "*Fakakta* theory. Great. I gotta tell that to my wife. She's Jewish, you know, but her parents were from Hungary so they didn't speak much Yiddish." He laughed at the thought. "A *fakakta* theory."

They walked into Oakley's office and Oakley placed his briefcase on a shelf. He picked up a few darts and started throwing them against a dartboard on the wall next to the door.

"Very relaxing," said Oakley, throwing a dart. "I always try for the seventeen and nineteen at the bottom. Even with the three, the expectation value is thirteen." He threw a second dart. "Better odds than trying for the twenty. See," he said, pointing to the top sector of the cart board, "bracketed by the one and five. Expectation value of less than nine." He threw the third dart.

"Forty-four," he announced to the cheering crowd. "My all-time record is one hundred and forty-eight. Two triple nineteens and a double seventeen. It's a game I made up to help me relax, unwind," he reminded Julian, as if excusing himself for his non-academic diversion. "I throw three darts and simply tally up the total. First of me to two hundred wins."

The last time a Management
Consultant earned his fee

He pointed to the dart board. "See . . . seventeen plus thirteen plus double seven. Forty-four." Oakley had taught himself to play tennis and had made the varsity tennis team in high school. He was competitive, even if the two competitors were both himself.

His office typified that of the creative academic intellect. Shelves were full of journals, books, reprints of articles, and manuscripts, and his desk was piled high with journals. Nonetheless, he could finger through the journals to find any article that he might need. A framed photo of Vivian hugging him taken by Allan at Larry Toy's restaurant had found a clear area on which to reside on one corner of the desk.

While Oakley was amusing himself, Julian had placed his large manila envelope on Oakley's desk and had opened it up. He pulled out a notebook and handed it to Oakley, now seated in the swivel chair behind his desk. The notebook contained hundreds of sheets of computer printout. Julian reclined in a wooden frame chair meant for student office-hour visitors.

"First," said Julian, "the 7044 has much larger working memory than the 704 you used. I saved a lot of timing by using arrays with larger dimensions than having to do a simulation, store the data on magnetic disk, reinitiate the variables, and start over. Then when I need to do comparisons, I don't have to read the data off the disk. And this way I avoid read errors, although they're rare."

Oakley was aware of the advanced computing abilities of the 7044, confirmed by their telephone conversations, but he let Julian continue as he looked through the sheets of computer printout in Julian's notebook.

"It also permits more nested DO LOOPS so I can use those variables of more dimensions that you could with the 704," said Julian, as if speaking of a child of whom he was proud, but a bit apprehensive about Oakley's response to his work. "I can store simultaneously the number of cards played, the number of hands played, the number of cards of each value played, the number of unseen cards that have been dealt, and so on. I don't have to approximate any of these parameters. If I wanted to, I could even distinguish the suits of the cards. But this is blackjack and not poker or gin."

"Perhaps someday variations in the game, or side bets, may involve colors and suits," observed Oakley. He remembered fondly the three-card flush he had announced to the dealer at Barney's.

"The code you wrote for your five-count and ten-count strategy is interesting, what you call the 'arbitrary subset' approach," said Julian, "but with the 7044 I didn't have to make approximations so I've revised some of it." He pointed to one of the computer printouts at the beginning of the notebook. "See? The results speak for themselves."

Oakley continued to leaf through the notebook of computer printout Julian had placed on his desk. Julian continued, confident but still apprehensive about Oakley's response to his work.

"I'm using essentially the same methods and techniques that you developed and used to work out both the five-count and your more sophisticated and quite powerful ten-count strategy systems," Julian said. He wanted to make sure that Oakley knew that. "The ultimate results will be more complicated, but hopefully with some simplification can generate something easily memorized."

Oakley knew that had been the plan. He politely raised his hand to silence Julian as he continued examining the computer printout.

Julian pointed to the page that Oakley was reading. "As you suggested, I removed one Ace from the deck, and these ten tables show the resulting player's advantages, one table for each of the dealer's possible ten up-cards," he said. "Then I removed two Aces from the deck, and generated another ten tables. And I did the same thing, removing from one to four of every card, and then the sixteen ten-value cards." This was what Oakley had begun doing his honeymoon night, but he was using a pencil and a piece of paper.

"So we have ten times fifty-two or five hundred and twenty tables," said Julian. "Each with ten entries for splitting and fifty-five entries both for doubling down and drawing or standing, one for each of the fifty-five possible player's two-card hands, including pairs. That's a total of one hundred and twenty entries in each of five hundred and twenty tables. More than sixty thousand probabilities to calculate by a Monte Carlo simulation. Actually, it's not quite that large because the dealer has a card. If he or she has a six showing, for example, I can only pull out the three remaining sixes, so it's really five hundred and ten tables. And you've got to include the possibility of multiple splitting of pairs." He waited for Oakley's response. "I brought all the tables along," Julian crowed, slapping the notebook. "It took the 7044 about a half an hour to an hour."

"The Four Horsemen of Aberdeen estimated they spent four thousand hours calculating their strategy," said Oakley. "To have performed these calculations would have taken fifty-one times as long, more than two hundred thousand hours. The basic strategy doesn't account for the cards that have been removed from the deck."

"Gad. Working six hours a day, three hundred days a year, that would have taken . . .," said Julian, pausing for a second to perform the mental arithmetic, "more than one hundred years."

"They wouldn't have made much of a dent even by now," concluded Oakley. "They would have needed all the king's horses and all the king's men, and they still wouldn't have been able to . . . whatever." Oakley failed to find a statistically relevant metaphor for Humpty Dumpty, but Julian understood.

"If you want to examine the use of multiple decks, you have that many more tables," Julian said, looking perhaps for an extension of a non-existent contract with Oakley. Actually, he just wanted to let Oakley know he was eager to continue the work and enjoyed working with him, not only because they shared a common interest but also because Oakley was a rare item, a mathematician celebrity, even if he was a lousy apprentice poet.

"Multiple decks, right," agreed Oakley.

"Here." Julian flipped through several hundred pages of printout. "I also calculated the dealer's probabilities of getting standing hands or busting not just for the entire deck but as the cards are dealt," said Julian, pointing to a particular page.

"Good," said Oakley.

"I did the programming, now it's your time to use the data," noted Julian, stretching out his arms to take a brief break from his presentation. "And to put it in a format that can be used in the casino."

"Good," said Oakley, relishing the creative challenge. "Let me tell you what I've been doing. My motivation for developing the five-count strategy system was my finding that removing the fives from the deck provided the greatest advantage to the player. In developing the ten-count strategy, I was aware of the crucial effect of the tens in affecting the advantages to the player. Now I follow what Harvey Dubner did."

"That fellow who contacted you after the convention?"

"Yeah. He's done some very good work."

Oakley opened his desk drawer and pulled out an index card.

"Here's a 3 by 5 index card he sent me with his results. He called the twos through sixes the 'lows' and the ten-value cards and Aces the 'highs.' In other words, although removing the fives from the deck provided the greatest advantage to the player, now we group the twos, threes, fours, and sixes in the same class. Similarly, now we group the Aces in the same class as the tens. They're also important. Dubner kept track of the number of each type of card that was played and seen and then subtracted the number of lows that were left to be played from the number of highs left to be played. It's the difference between the two types of cards that remain that he found to be significant. It's the basis of what Harvey Dubner called the 'Basic Hi-Lo Strategy.'"

"Nice. In effect, he said that all the lows had the same effect and all the highs had the same effect," observed Julian.

"That's exactly right," responded Oakley. "Now, can we quantify this rather than subtracting the number of each class of card dealt? I said sure, it's easier to just assign values to the cards. I assign a point value of +1 to the twos through sixes as they are played and seen by the player and one of −1 to the ten-value cards and Aces as they are played and seen. What I call the point total or total point count, or just point count for short, is the arithmetic sum of the point values of the cards that were played and seen, the total point value of the cards. In the end, you remember only one number, the sum."

"Instead of two numbers, the number of highs and lows remaining," noted Julian. "Easier to do."

"Right. It's as simple as the Goren point count in bridge."

"What about the cards that are played and not seen," asked Julian.

"If you don't see them, you can't include them in the count. They're considered as not being played," responded Oakley.

"And, what about the sevens, eights, and nines?" Julian asked.

"Even my calculations with the 704 showed that they have small effects, so I gave them a point value of zero. Harvey in effect did the same by excluding them from the categories of 'highs' and 'lows.'"

Julian flipped through his notebook. He pointed to one of the computer printout pages. "The 7044 calculations agree. They're relatively neutral in affecting the player's advantages. It's easier just to assign a point value of zero."

"Yeah. The small increase in sensitivity isn't worth the cerebral labor."

"Adding up the plus ones, minus ones, zeroes, and so on is a very nice concise way of quantifying the remaining cards," noted Julian. "We're essentially reducing the game to a two-type-of-card game, two classes of cards. You and Harvey are very clever, Ed."

"Harvey thanks you. I thank you. My teachers thank you. My thesis committee thanks you. I guess my wife thanks you," said Oakley, lightening up from his professorial chores.

"You're welcome," said Julian, perfunctorily. He was gracious, but he was now thinking of how this scheme could organize the multitude of his calculations and make it more palatable to Oakley. "Each of my card removal tables corresponds to zero, plus one, minus one, or a multiple."

He lifted his notebook from Oakley's desk and let it drop onto the desk, creating an audible thud. "Instead of considering each of my five hundred and ten tables separately, we can essentially combine them based on their point count," said an obviously relieved Julian. "This greatly simplifies any analysis."

He stood up and raised his forefinger to the air. "Mr. Braun is happy to announce that you won't need an electronic computing machine strapped to your back to apply the system in the casino. He suggests that you simply add and subtract plus and minus ones."

Oakley laughed at Julian's out-of-character antic. He had sensed Julian's apprehension and was now happy that Julian was relieved of that burden.

Oakley turned to two computer printout sheets in Julian's notebook. "Exactly. Here are two of your tables, one generated by removing five ten-value cards and this one generated by removing three of the deuces. The point count is minus five plus two, for a total of −3. The same point count results if the dealer had dealt out, say, twelve ten-value cards and nine of the low-value cards. To use your data, we only need know that the point count is −3, not which particular cards have been played. That would be nice to keep track of, an ultimate, ultimate system, but the typical player would not tolerate the challenge. I wouldn't."

"Nor I," agreed Julian.

Julian knew well that, as good as his results were, the ability to use them in the casino, put them in a casino-friendly "format," make it palatable as I said, was one of Oakley's major motivations, if not his single most important one after, of course, developing the mathematics of the system. This concept of the point count had made using his voluminous calculations practical, if still requiring some mental arithmetic. The theoretician in him, unfortunately, led him to suggest even greater complexity.

"You would get a greater discrimination among decks if you assigned different point values to some of the cards," offered Julian. In thinking about the plus-minus scheme, he had considered such an option.

Oakley had thought of this himself. "Sure," he agreed. "The five-count strategy was based on their absence providing the best advantage to the player of any card, so I could assign, say, +2 to the fives, and because the absence of Aces is disadvantageous I could assign −2 to them.

"But, will this really generate an improved discrimination in the quality of the deck? I don't think so. First, they burn a card, which you can't see to include in the count. That by itself introduces an uncertainty. If it's a five or an Ace, the error could be plus or minus two instead of plus or minus one. The mistakes you'll inevitably be making in counting the cards introduces more uncertainty in the count if the absolute value of some are increased, and making the system more complex will tire the player and increase the frequency of those errors."

"But if the errors are random and relatively few, it shouldn't affect the player's expectation values."

"Yeah. In the long run, but not necessarily in the short run, right? So I've concluded that any marginal improvement in discriminating, as I said, differences in the quality of decks might not be realized. In short, it's what I think is just an unjustifiably more difficult calculation," he concluded.

"Which you want to do in your head," noted Julian.

"In the heat of the battle, so to speak," agreed Oakley. "You want to do it quickly, effortlessly really. Unless you want to sit there using a sheet of paper and pencil and tip the casino off that you're playing this system."

"Or wear an electronic computing machine on your back," added Julian. He repeated the witticism, happy that Oakley had rejected a more complex counting scheme. He could only have imagined the computing he might have been asked to perform. As much pleasure as it gave him, IBM, after all, wasn't paying him to run blackjack simulations. Julian knew how a think tank works, and IBM was to a large degree a think tank. The algorithms that he was developing and using to analyze blackjack might indeed be applicable to work that IBM might now or in the future consider relevant. IBM was a business, however, not a university where academics can pursue their research interests, no matter how unconventional, arcane, or even unpopular with colleagues they may be. At least, Julian had repeatedly reassured himself, his supervisors had never complained about his hobby.

While Julian was immersed in his thoughts, Oakley was immersed in his. Ever since he had played Baldwin's system, using that cheat sheet which the dealers and other players mocked, Oakley believed that the effect of removing

LEGENDARY "BLACKJACK BOB" CAN COUNT
UP TO 273,084,137,314,159,265 DECKS

various types of cards would change the advantage to the player. He thought about his comments on the drive home with Vivian, Allan, and Joan, about his dealing hands to himself on his honeymoon night, and about the hours at UCLA he had spent trying to devise a system by hand.

Oakley took off his glasses, walked over to his blackboard, picked up a piece of chalk, and wrote down a formula. He addressed the classroom.

"Now, Julian, in what I call the simple point count system I used the point count only as a rough guide to betting and as an approximate indicator to the player of doubling down and splitting more frequently and of standing more frequently when the point count was large and positive, and to do the opposite when it's negative," said Oakley. "Other than that the system follows the guidelines for playing my recalculation of Baldwin's group's system, what I call the 'basic strategy.' In it, we only consider the dealer's up-card and the player's two cards."

"With the more precise results from your 7044, we're now justified in using the point count to calculate an index to guide the strategy decisions. It's what I call the advanced or complete point count strategy," Oakley announced. "Borrowing Harvey's 'Hi-Lo' terminology, I also call it the 'high-low strategy.'"

"The 'high-low strategy.' I like that," said Julian. "The index?"

"Yeah, it's just a small additional bit of mental math needed. I divide the point count by the number of cards left in the deck and multiply by 100," Oakley lectured. "See," he said pointing to the equation he had written on the blackboard. "The high-low index, I, equals 100 c/n, where c is the point count and n is the number of cards left to be played. I multiply by 100 to provide an integer result rather than a fraction. If you don't see a card, you don't include it in either the point count or the number of cards that have been played. The values range most frequently between -25 to $+25$."

"Good. Catchy," said Julian. He let the phrase roll off of his tongue. "The high-low index."

"A deck of $n = 20$ cards with a point count of +2 is richer, so to speak, in cards favoring the player than a deck with $n = 40$ cards left with the same point count," continued Oakley. "Conversely, a deck of 20 cards with a point count of -2 is poorer in cards favoring the player than one with 40 cards left. So you need to remember two numbers, the point count and the number of cards that have been played and viewed."

"Harvey considered this in his hi-lo system," noted Julian, pointing to Dubner's 3 × 5 index card.

"Yeah. It's very similar to Harvey's 'Hi-Lo Index,'" agreed Oakley, looking at the 3 × 5 index card. "As I said, he subtracted the number of low-value cards left in the deck from the number of high-value cards left in the deck, and then he divided as I do by the number of cards remaining to be played. I followed his lead. He called it the 'Hi-Lo Index' and used it to determine his bet size and hard drawing and standing strategy, but he didn't then multiply by 100. Without that you're left with decimal results and it becomes challenging to use I think."

"I tend to agree," said Julian.

"His calculation of the 'Hi-Lo index' and my calculation of the 'high-low index' give the same result. Let's say three ten-value cards and Aces and five low-value cards have been played. My point count would be -3 plus $+5$ or $+2$ and the high-low index would be $+2$ divided by 44 times 100. Harvey's Hi-Lo Index would be calculated as the original 20 high-value cards minus the 3 played or 17 high-value cards left in the deck, minus the number of low-value cards left in the deck, the original 20 minus the 5 played, or 15. And then subtracting the two gives $17 - 15$ or $+2$, just as I would calculate."

Julian rose and went to the blackboard. "Yeah, it's trivial to show they're equivalent except for the multiplication by 100," he said. "Here," he wrote as he spoke, "let H be the number of high-value cards played and L be the number of low-value cards played, and N be the number of cards in the complete deck, 52. Then you in effect calculate $L - H$ and divide by the number of cards left to be played and Harvey calculates $(N - H) - (N - L) = -H + L$ or the same $L - H$ and again divides by the number of cards left to be played.

"I like your method better. As you said, much easier. You can skip three steps counting the cards as they are played rather than calculating the number of cards of each type left to be played and then subtracting them. It has the same added complexity as your ten-count strategy. I don't know if he knew about the ten-count strategy when he was developing his Hi-Lo, but he's using the same way to calculate the index."

"The entire concept agrees with intuition," Oakley noted. You would stand more frequently if the index is positive, meaning you've got lots of high-value cards left in the deck, but take a hit more frequently if the index is negative, because you have a better probability of drawing a low-value card.

"It's similar to what I did with the ten-count strategy, where I calculate my ten richness ratio as the ratio of the non-tens to tens which have been played and viewed. But those values with their decimal points make the system cumbersome to use, challenging. Harvey probably saw the same thing with his Hi-Lo Index."

"You must love indices," noted Julian. "Your five-count strategy also calculated an index," said Julian, remembering Oakley's presentation at the mathematics meeting.

"Right, number of unseen cards divided by the number of unseen fives," said Oakley. "But this, Julian, this provides a more precise quantitative means of determining strategy decisions beyond the basic strategy and the five-count based on the quality of the remaining cards. You give me the player's hand and dealer's up-card and all the strategy decisions are determined by this index. And with your results, the guidance will be more precise. I can now create more precise strategy decision numbers for the player related to the high-low index. And I'll also be able to provide a guide, like I did in the ten-count strategy, for taking insurance. Dubner did the same. This is very, very good."

"This is more detailed than what Harvey proposed," added Julian, looking at Dubner's 3 × 5 index card. "He gave hard and fast rules for splitting, doubling down, and soft drawing and standing, and insurance, as if he were playing the basic strategy. He provided strategy decisions based on values of his Hi-Lo Index for only hard drawing and standing. Yeah, our results will be more detailed."

"And, as you say, with the 7044 more precise," added Julian.

Oakley sat in his swivel chair. Then he suddenly pumped his fist in the air in exultation. "Holy, holy, holy, Moses . . . Cheese . . . Cow . . . holy Toledo. Ohhh., Julian!" For an atheist, Oakley had a lot of gods.

Julian was pleased that Oakley was pleased. It was obvious that Oakley was much more, in fact, than simply pleased.

"The beauty of a 7044," said Julian, modestly. He was being more than modest, of course. He wanted to ensure that Oakley knew that he was an indispensable partner. Julian, despite his intellect, was somewhat of a recluse. He could hope that some of Oakley's attributes, a man with a vivacious wife, enviable academic position, and bright future might, if not further his career, at least spice up his life sitting at a desk at IBM.

Oakley sat in his swivel chair and leaned back. He stared at the wall. They had created a monster, if a mathematical rather than biological one. He got back up and walked to the blackboard. He sketched out a rectangular coordinate system and wrote "E" next to the ordinate scale and "I" next to the abscissa scale. He put a minus sign below the abscissa scale to the left of the origin and a positive sign below the scale to the right of the origin. He drew a continuous line, beginning as a concave upward curve in the third quadrant, straightening out as it approached and passed through the origin, and then continuing as a straight line at close to a 45-degree angle upwards into the first quadrant.

"If I made a graph of the expectation value for a one-dollar bet as a function of the high-low index, it would certainly be an increasing function, if you follow the strategy decisions," said Oakley pointing to the graph, in a capsule summary of their work.

Julian started to make a comment, but Oakley knew what was coming. He raised his hand. "I said 'if you follow the strategy decisions.' The index will be negative as frequently as it will be positive, but the player will be changing the strategy. You've got to follow the system."

"Ah, yeah," said Julian.

"And it provides a guide to betting," Oakley added, in a trance. You bet one-half of the high-low index."

"That's very interesting," said Julian. He pointed to the formula Oakley had written on the blackboard. "As the deck gets richer in cards favoring the player, he or she should increase the bet, and this provides the formula," Julian joyfully announced, becoming unable to control his enthusiasm at their mutual tutorial.

"Fortune's formula," added Oakley. "Better than the betting concept I had for the simple point count where, in its simplest form, you just bet the total point count, or your minimum bet if the count is zero or negative."

Julian smiled, remembering that phrase as part of the title of Oakley's paper to the American Mathematical Society meeting.

"Lower your bet to a minimum for low or negative values of the index, reducing your expected loss, and increase it when the index is positive, increasing your expected winnings," instructed Oakley. "This follows what Harvey suggested in his Hi-Lo system. He suggested $1 bets when the hi-lo index was small or negative, $5 bets when the index was between 0.05 and 0.10, $10 bets when the index was between 0.10 and 0.20, and $20 bets when the index was greater than 0.20."

"So he basically said bet half of the upper boundary of the interval of the Hi-Lo Index, multiplied by 100," concluded Julian. "Again, our results will be more detailed."

Oakley pointed to the curved line he had drawn on the graph, and drew a second curved line. It began from the origin but curved above the first line he had drawn.

"Changing your bets with changing the high-low index increases the expectation values even further. I developed a rough scheme on increasing the bet with the ten-count strategy system, but Baldwin's system, my basic strategy, the five-count, and even the simple point count essentially said just to raise the bet when the deck became favorable. Basically raise it as high as possible, except it may tip off the house when you consistently win a large fraction of those large bets," he said, finishing the lecture to his one-man class on the subject of the high-low index. He sat down, slapping his hands together to dissipate the chalk dust.

Julian walked over to the dart board against the wall and pulled out Oakley's darts. He walked back about ten feet and threw the darts one by one trying for the bull's eye. The darts stuck on the three, eleven, and eight sectors.

"A twenty-two," said Oakley. "Not bad for a first try, Julian. With a little more practice you could be named rookie of the year." He pointed to his desk. "I want to show you something."

Now it was time for Oakley's show and tell. He carefully pulled a writing tablet out from the pile of books on his desk and placed it beside Julian's notebook. Julian stared in amazement, not only that Oakley could find the tablet but that the pile of books did not come toppling down.

"These are the tables in the format I devised for my recalculation of Baldwin's system using the 704, what I call the basic strategy, and for my ten-count strategy. The possible dealer's up-cards from two through Ace are along the top row and the player's possible two-card totals starting from twelve up to nineteen are along the left-hand column," said Oakley, pointing to the top page of the tablet. The entries in the matrix direct the player to draw or stand. The cell being darkened means that's the minimum standing total," said Oakley in lecture mode.

"The Four Horsemen provided strategy charts in their book with sections on drawing and standing, doubling down, and splitting pairs," observed Julian. "Yours is similar."

"Yeah, I know," said Oakley. Here, see, this is the hard drawing and standing table for the basic strategy. With the player holding a 13 or more against the dealer's 3 you stand. If you have less than 13, you take a card. Same holding 12 against her 3."

"The dreaded 12 and 13," moaned Julian.

"And 14 and 15 and 16," Oakley added, "the heinous, malevolent, marginal hands," Oakley continued, apparently ascribing free will to the cards. "In the comparable tables I'll create for the complete point count system, I'll put the values of the high-low index to guide the standing decision in the cells where the decision isn't clear-cut and obvious, that is, where the value would not be either a very large positive or negative number. If the player's high-low index exceeds the value in the cell, he holds. If it's lower than the value in the cell, he takes a card."

"This is very clever, Ed."

"I thank you. It's a very concise summary of your type of work. If that means clever, so be it. The ten-count strategy is also powerful, but the ratio of non-tens to tens is always a small decimal fraction, and memorizing and recalling them can produce migraines."

"Harvey Dubner presented his results more or less verbally," continued Oakley. "I think presenting them in this tabular form will be easier for the player to use."

He turned over the top sheet to reveal a second. He let Julian stare at it for a moment. "This is a similarly laid out table for hard doubling, and here's the table for soft doubling. And," he said, turning to a fourth page, "here's the basic strategy table for splitting. In each case the format is the same, the dealer's up-card along the top row and the relevant player's cards along the left column. That's the two-card total for hard doubling, combination of Ace and another card for soft doubling, and the various pairs for the pair splitting table. Values of the high-low index determining the decision will appear in these cells. The table for soft drawing and standing is much simpler, with just a couple decisions not clear-cut."

"I think you'll find that most of the strategy decisions will be the same as in the less sophisticated systems, Baldwin's, your basic strategy, and your five-count strategy," said Julian with the experience of one who had performed the most extensive programming of the game.

"It's the marginal, difficult decisions that will be decided, based on your results, by the complete point count system," said Oakley, almost bragging, although a large portion of the credit was due, as he knew, to Julian. "Yeah," said Julian pointing to an entry in Oakley's table for holding and drawing. "Here, for example, if you've got a total of 12 against the dealer's 6, or a 13 against her 2."

"Or double down with a 9 against her 3, or an 11 against her Ace," said Oakley referring to his second page. "The basic strategy says double down holding an 11 and, depending on the dealer's up-card, usually a 10 and sometimes a 9. The index will uncover other favorable situations."

"This is a much more powerful system."

"Yeah. These are difficult decisions for the player." Oakley showed Julian the fourth page again. "Here, the splitting table for the basic strategy says to split fours against a five. I'm sure the complete point count will provide guides to split against her 4 and 6, too. It'll provide guides for standing or drawing, doubling down, and splitting that a player using the basic strategy would never consider."

"Even splitting tens, perhaps," suggested Julian.

"The basic strategy says never split tens," noted Oakley. "But with a high value of the index, indicating many tens and Aces remain in the deck, that might be a good move. The ten-count strategy uncovered such cases."

"But you never tried it out in the casinos," remarked Julian.

"Nope. I never tried it out in the casinos," said Oakley. "But that's going to change."

"From your mouth to God's ears," said Julian, repeating one of his mother's favorite phrases. "But you're right, winning these marginal cases should markedly increase the player's advantage," Julian summarized.

The super-sensitive sense of the humor of the brilliant mind was manifested as Oakley laughed at what 99.999% of the population would consider an extremely dry, humorless statement.

"That is the real advantage of the complete point count system," chuckled Oakley. "With the 704, I found an overall player's advantage of 0.12% playing my recalculated basic strategy, and up to a 3.6% advantage playing the five-count, when all the fives have been played. I recalculated Baldwin's strategy and found a 0.09% advantage to the player. Using your results, I wouldn't be surprised to find advantages of 5 to 10% or more using the complete point count system."

"Or 15% or 20%," laughed Julian.

"Hey, you want to get some tea or coffee?" asked Oakley. They both needed a break.

"Sure." They walked out of Oakley's office down the hall to the faculty lounge.

"If this works, Julian –"

"The 7044 hasn't lied to me yet. Oh, it'll work."

"When this works, Julian, we're going to publish it."

"Yeah."

"Maybe, maybe not in the probability journals. Maybe in a book."

"Wow. That's not the traditional mode."

"I have things to prove to some people."

"Be that as it may."

"Then, I'm sure when 7044's become widely available others will hop on board and use more complicated point value assignments. But it won't markedly increase the actual casino performance."

"Braun and Oakley," Oakley wrote in the sky.

"No, you started it at the AMS meeting. It'll be Oakley and Braun. And we should really include Dubner."

"Have you tested your results?" asked Oakley. It had only been a few weeks since the American Mathematical Society meeting, and Oakley remained sensitive to the criticism of the skeptics, the press, some mathematicians, and the public. He knew that Julian, having been in attendance, would be aware of this need.

"I let the computer deal nine billion hands," answered Julian. He could tell from Oakley's lowering his head that the answer wasn't sufficient. "And I've dealt myself some decks."

Having arrived in the lounge, Oakley poured hot water into a cup. "Coffee or tea, Julian?" He pointed to a group of tea boxes.

"Green tea looks good," said Julian, pouring hot water into a cup for himself and pulling out a tea bag from one of the boxes.

"No, I mean, at the casinos?" Oakley knew Julian was dodging the question. He knew the situation as well as Oakley.

"I was at the meeting, too," said Julian, admitting the obvious. "They're brutalizing you. But I really don't gamble. It's . . . I don't like it, the atmosphere. The fear of losing my modest life's savings." He hoped that Oakley would accept his excuses. They sat down at a table.

"Okay," said Oakley. "Look. Manny Kimmel is coming up from New York in two weeks. He's having dinner at the house. I think he wants me to teach him the ten-count strategy. Hopefully we'll have finished the analysis of the complete point count by then. I'd like you to be there. Help me feel him out. If he can learn it and win in the casinos, it'll silence the skeptics for good. Can you come back?"

"You chose Manny Kimmel?" said Julian, relieved that he would not be chosen Vice-President of Casino Testing.

"Consider it a final interview," said Oakley. "They all have money. But he's a multimillionaire, and I wanted to involve someone who wouldn't be hurt if they lost their entire bankroll. He's also a big blackjack player, so I think he's sold in advance on our system and won't give us trouble if things go bad. Some of the others had similar traits, but I think he's pretty savvy when it comes to spotting cheating dealers. That's really our only obstacle."

"Cheating," said a happy Julian. "That's the one thing I can't code." He finally could crack a joke, although it seemed to be his only joke, being similar to the one he made with Nancy about coding stupid moves at chess.

"What? You can't?" said Oakley.

They both laughed. No matter how much faith they had in their mathematics, both, particularly Oakley, wanted the system tested in the casino. He needed to silence his detractors and he needed to dispel his own doubts, no matter how irrational they were.

12

Forbidden Fruit

Boston University, Boston College, Harvard, MIT, Northeastern, Tufts, the Berklee School of Music, renowned for its jazz program, Radcliffe, Smith, Brandeis, Amherst, Simmons, Wellesley, U Mass, and other colleges and universities make Boston, Cambridge, and the Boston area their home. Even with all the Irish alcoholics lying in the streets, it was always a great place to go to college.

The tens of thousands of professors, staff, and students, though, have a demand for housing, and, although many students live in the dorms or fraternities during their freshman and perhaps sophomore years, by their upperclass years many look for apartments or houses to rent. For a newly arrived faculty member, especially if married, finding congenial, affordable, and relatively nearby housing can be challenging. College and university housing offices can help find nearby apartments, but for Vivian and Oakley, the options were not acceptable.

They were able, however, to find a nice, small rental house in Brookline, a western suburb of Boston just across the Charles River and a bit west from Cambridge and MIT. It was a quick ride on the Metropolitan Transit Authority, the MTA of the George O'Brien themed celebrated song by the Kingston Trio, to the Massachusetts Avenue stop and then either a bus ride or a hitchhike across the river. If Oakley had been a bike rider, he could have made the trip on Beacon Street, past Fenway Park, and across the Harvard Bridge on Massachusetts Avenue directly into campus in 15 minutes.

Hitchhiking across the Harvard Bridge was not fraught with the usual hitchhiking anxiety of not succeeding and, in this case, missing classes. The scads of students standing there, thirty at a time, thumbs in the air, posed no crime threat to drivers, and the students had no reason to fear any accommodating drivers. Crime just didn't exist hitching rides across the Charles River on the Harvard Bridge.

The house the Thorps had rented had an enclosed garage, where Oakley felt his T-Bird would be safe not only from youthful vandalism but also from the weather, his having been warned about the ferocity of nor'easters. One dent was enough.

Their oval dining room table was covered with white linen this special evening. Coffee cups and saucers, a bowl of fruit, a box of chocolates, a vase of flowers, linen napkins, and spoons lay on the table, around which seven chairs sat. A breakfront cabinet, bought from a used furniture store, judging from the fine scratches on its cabinet doors, stood against the wall. Almost all of the furnishings in their home were rented or bought used.

The little party to both entertain and interrogate Manny Kimmel was jovial enough. Oakley, as the host, sat in the chair at one end of the table. Julian and Nancy sat to his left. When Julian had driven up in his rented car with Nancy, Oakley and Vivian had been a bit surprised. They had no idea that Julian, he the bookish type, had a girlfriend, and they certainly hadn't been forewarned that she was coming along. That crisis was minor, Oakley and Vivian quickly discovering that Nancy was an educated, respectable woman, and, more importantly, that Julian and Nancy hadn't planned on making the Oakley home their hotel. It had gotten warm enough in Massachusetts, with the temperatures predicted to be in the high 40s, that Nancy and Julian had planned a trip to Cape Cod.

Manny Kimmel and the women he introduced as his nieces, Madeleine and Renee, sat to Oakley's right. His entry was, shall we say, a bit more ostentatious than that of Julian and Nancy. Driving up in a large, midnight blue Cadillac is one thing. The car door opening for the exit of two ravishing, voluptuous blonde women in their 20s who were wearing chartreuse stiletto heels and revealing dresses under their mink coats was another. It's a good thing that the driveway up to the Oakley's house was partially shielded by the house itself. Residents of this area of Brookline weren't accustomed to seeing a large Cadillac spouting out two sirens, nieces or not.

Manny Kimmel was in his sixties, white-haired and small of stature. Sitting between his nieces in the car, he had been hidden from view. A prominent diamond ring weighed down one of his fingers. The *de rigueur* gold chains draped his neck.

Dinner had been served and Nancy had helped Vivian clear the table, a service that was much appreciated. Vivian stood behind Madeleine ready to serve, holding a coffee pot and a tea pot as Julian and Nancy sipped from their cups. Oakley sampled a piece of cantaloupe from the fruit dish. Most of the pieces of fruit in the fruit bowl and most of the pieces of boxed chocolate would be consumed during and after coffee and tea.

"Thank you, Vivian," said Nancy.

"Coffee or tea, Madeleine?" asked the hostess.

"Could I have a little apricot brandy, please?" asked Madeleine.

"My niece, see, she likes fruit," explained Manny.

"Lots of vitamins," Madeleine added.

Vivian put the coffee pot and tea pot down on the table and walked over to the breakfront cabinet. She pulled out a liquor bottle.

"Well, special treats for special guests," she said.

Manny gestured toward Vivian.

"You're a lucky man, Thorp," he said.

"There's no such thing as luck," Julian corrected him.

"See, doctor," said Manny. "That's where we differ. There is luck. And with me bankrolling your little experiment, I expect a lot of luck."

He turned to Nancy.

"You believe in luck, Nancy? You know, fate?"

Vivian returned to the table with the liquor bottle and a schnapps glass.

"We didn't have apricot brandy," she apologized to Madeleine. "Is peach brandy okay?"

"Sure," Madeleine giggled.

Vivian poured a little of the brandy into the schnapps glass. Madeleine watched her pour.

"A double, please," she said.

"I think Mr. Braun would agree that probability is important," Oakley replied to Manny. "But luck in the sense of an external force of some sort is not realistic. It has no place in rational inquiry."

This was not the kind of discussion to which Madeleine or Renee was accustomed. Madeleine was closely monitoring the level of booze in her glass as Vivian added what she was guessing would be equivalent to a second shot. Renee rolled her eyes in disbelief at the conversation.

Vivian, finished with Madeleine's glass, walked over to Renee with the coffee and tea pots.

"Gad," Renee said, turning her attention from the conversation to the liquid offering. "You got a Pabst or a Bud?" she asked before Vivian could offer the coffee or tea.

"She ain't no health freak like her cousin," said Manny. "No sweet tooth, either."

"I think so," answered Vivian, leaving the room for the kitchen.

"You got a regular speakeasy here, Thorp," said Manny.

"Well, every once in a while, Mr. Kimmel." He stared at the kitchen. This wasn't the kind of treatment he had expected for his wife.

"What other kinds of ways can they cheat you?" Julian asked.

Vivian returned with a bottle and a glass.

"Here's a cold Bud and a glass, Renee," she said, delivering the goods.

"Oh, goody," responded the non-health freak. Manny's supposed niece couldn't wait, and guzzled the Bud from the can. This didn't bother Manny, having no doubt seen it before.

"Coffee for me," he said to Vivian. "No sugar."

Vivian picked up the coffee pot from the table and poured some coffee into Manny's cup. She sat down, audibly, as if letting gravity do its thing unhindered. She was sending a message to Oakley.

"Dealing the second card," Manny answered Julian's question. "They peek, see, sometimes using the reflection in a pinky ring. If they don't want the card for you or themselves they deal the second. So, you check their fingers. Pinky ring? You run."

"Marked cards?" asked Oakley. This was a business meeting, and although he was aware of Vivian's playing the unintended barmaid, they had to see if Manny possessed the casino street knowledge that he claimed.

"No," said Manny, "they don't do that no more. People who say that are full of hot air. The decks are clean. See, the regulators, they're all over the house. The decks, they unseal them at the table and spread 'em out." He made a sweeping motion with his hand, simulating this activity. "You can examine them. But they do mark 'em during the play. They fold back the edge a little, see? Undetectable."

"False shuffle?" asked Julian.

"Yeah. That's a real problem," answered Manny. "The good ones, the mechanics, they shuffle and keep the order."

"Mrs. Thorp," Renee called. "You got another Bud? That hit the spot."

Vivian had been sitting for barely two minutes. Under the table she had slipped her feet out of her flats for relief. She stared at Oakley before jovially answering.

"Sure," she said. "Coming up."

Vivian rose out of her seat and walked into the kitchen. Oakley took a piece of fruit. He was trying to show everyone that everything was okay with Vivian, to the extent that his stomach wasn't churning in empathy the way it in fact was.

"Mechanics?" asked Julian, unfamiliar with the term.

"Card sharks," Oakley answered.

"Right," agreed Manny. "They're called mechanics. See, they're professional cheaters. On call from a bunch of houses."

Vivian returned with a second can of Bud. As she placed it in front of Renee, Madeleine took a piece of chocolate from the box of chocolates and bit into it.

"Ewwww!" exclaimed Madeleine. She turned her nose inside out and swung her head to the right, apparently in some kind of survival mode, then placed the uneaten half of the piece of chocolate back into the box of chocolates. In response to this display, Julian and Nancy looked at each other in disbelief and Vivian stared at Madeleine. Vivian sat down, although she wouldn't have been blamed if she had simply fainted. Julian and Nancy were now quite sure that Madeleine had not been raised in a household conforming to upper-crust British etiquette.

"Oy," said Nancy under her breath.

"You can always tell 'em, the mechanics," Manny continued, oblivious to the candy game and body language. "They come in any time, not just the shift breaks. And they don't wear the house uniform."

Madeleine licked her lips. Vivian had been sitting down now for nearly thirty seconds. "Mrs. Thorp, you got a Snickers bar maybe? Or some M and M's?" Fine chocolates weren't her style, it seems.

Vivian got up and walked to the kitchen. This time she walked slowly.

"That reminds me," said Manny. "The uniform. You gotta hide you doing this system stuff. Be a regular bullshit artist. Pretend you're ski bunnies or something, you know, there for skiing. Wear a parka. You know, fit in. Or booze hounds. For the free drinks. Slobber it on you, so you reek of the stuff."

"Vivian and I are musicians," Oakley said. He wanted to make sure to include his wife, and spoke the words loudly enough so she, still scrounging for some chocolate in the kitchen, could hear it.

"Yeah, that's good," responded Manny. "You came to hear the music or something."

"Well, how do you protect yourself against the false shuffle?" persevered Julian. He was aware of the impending nuptial crisis, but he hadn't driven in from Chicago to watch a parade of beer and chocolates.

"Yeah, sorry, a problem. But it's hard to keep the meld if you shuffle more than twice. And you gotta be sure, see, that they cut the deck first, shuffle each part, and then put the deck together and shuffle again. That pretty much wipes out the melds unless the guy's really good."

Oakley looked toward the kitchen. Either Vivian was sitting somewhere taking a break or was digging very deeply in drawers looking for those Snickers or M and M's.

"Bottom dealing?" he asked.

"No," answered Manny. "See, they burn a card but flip it first. So, see, the bottom card is turned over. Too obvious, anyway. Guy turns his palm, looks at the bottom card. Nah. That don't happen no more. You're not going to have to worry about that when you're playing. You and the wife are musicians. That's a good cover."

"What do you mean?" asked Oakley.

"You don't have to worry about that," repeated Manny. "They ain't interested in cheating kid musicians. Pretend you're there to get a gig."

"What? I'm not gambling," stated Oakley.

"Sure you are. I'm providing the cash. A hundred big ones. You're gambling and doubling it. That's it."

"I thought you wanted me to teach you the system. Look, I've got to prove to all these doubters that the system works, but you've got to play." Oakley was starting to get an empty feeling in his stomach. "And I'm uncomfortable with such sums," he added. "Ten thousand is enough."

"No no no no no no no," said Manny in a near stammer. "They know who I am. And I can't do your math stuff. My nephew Seymour, he's a math genius, too. He could do it, but not me."

He stared at Oakley. "You're going to do this. Me and some of my best friends, we provide the cash and take 90% of the profit. It's a business deal. Ten thousand? That's okay. You're doubling it. We make the profit, and you try out your system."

Manny Kimmel, born Emmanuel Kimmel, was a bookie and gangster in New Jersey and New York dating from the Depression. Unknown to Oakley, he had been considered a "notable underworld figure" for decades. He was considered to have run the largest handle horseracing book in a crime-ridden, book-happy, corrupt New York City. He worked with the single biggest racketeer and bootlegger in Newark, Abner "Longy" Zwillman, a feared Prohibition gangster, leasing his garages to Zwillman for storage of liquor during Prohibition. So well-known were his transgressions that the FBI tracked him for his relationships with the mob and eventually persuaded him to testify against Zwillman, his former associate.

He and his wife Pauline (Polk) had founded the Kinney Parking Company, a prominent national chain of parking lots and garages. The name was suggested by its first location having been on Kinney Street in Newark.

That Manny had adopted this name to hide his Jewish identity was the *modus operandi* of American Jews from the turn of the twentieth century to its end, a "time-honored tradition" of American Jews anglicizing both their given and business names to hide their Jewish identity. This was done not only to avoid anti-Semitic persecution and discrimination but also to fulfill a desire to be accepted and assimilated into American society. Although they still read newspapers published in the Yiddish of Eastern Europe, supporting dozens across the country, the pride of first generation-born American Jews was to learn and become fluent in English.

Yet, they still faced persecution and discrimination. Most professions were to some degree closed to Jews, including architecture, medicine, law, and veterinary practice. The famous Cornell University veterinary school, for example, through the early 1940s, allowed only one Jew in an entering class. Jewish animal lovers reacted by founding a veterinary school at what was to become Middlesex University in Waltham, Massachusetts, with a faculty of Jews who had fled Germany.

Colleges and universities, including most of the elite universities, even until the 1980s, had such Jewish "quotas," severely restricting the number of Jewish students they would admit, previous academic success and future academic promise being irrelevant. With programs in medicine, pharmacy, podiatry, and veterinary medicine, as well as the liberal arts, Middlesex was the only university in the United States at which the medical and veterinary schools did not impose a quota on Jews. The American Medical Association refused to grant accreditation to its med school, which its non-Jewish founder partially attributed to institutional anti-Semitism. After the founder's death, Middlesex evolved into the predominantly Jewish Brandeis University in 1947.

Socially, Jews were isolated. The Jews who entered college would form their own fraternities and sororities, the Sigma Chi's and Tri Delt's not rushing Jews for members. Out of college, with admission to social and athletic Clubs denied, the Jews formed their own. Julian Braun's parents belonged to the Covenant Club, formed by Eastern European Jews. Chicago Jews of German descent formed and joined the Standard Club.

One of the few fields open to Jews was physics, that not being a calling of most non-Jews. With the field open, most of the breathtaking developments of the 20th century were made by Jews. The names of Jews, Hans Bethe, Albert Einstein, Wolfgang Pauli, Richard Feynman, Albert Michelson, Max Born, James Franck, Emilio Segre, Arno Penzias, Murray Gell-Mann, Leon Lederman, Niels Bohr, Steven Weinberg, and others, are inscribed in the listings of Nobel Prize winners in gross disproportionality to their numbers. J. Robert Oppenheimer, the genius who directed in the Manhattan Project, which contained the greatest assemblage of physics intellects ever assembled, was Jewish. This success aside, Jews were denied entry into most other American professions.

Jews, indeed, created their own American industries where they would not suffer discrimination in opportunity and hiring. Three of the five American-born entertainment industries were of Jewish origin. New York City Jewish veterans of vaudeville performed in the resorts of the Catskill Mountains, and there, in the Borscht Belt, stand-up comedy was born. Those with a musical and theatrical bent founded musical theatre, an outgrowth of Eastern European Yiddish Theatre and the source of the popular music canon known as the "great American songbook." The motion picture industry was founded by Jews who moved to southern California. Of the five American-born entertainment industries, only vaudeville and jazz, created in New Orleans by blacks, were not of Jewish origin.

That discrimination in the professions and colleges continued after World War II. The 500,000 American Jews, an absolutely remarkable half of all Jewish males between the ages of 18 and 50, who served, fought, and became injured or died in the war returned to a bigoted America. Many, facing this discouraging fact, opted to devote themselves to the creation of a Jewish homeland. Israel was built on the intellect and industry of transplanted Jews. In a significant way, American bigotry led to its creation, existence, and flourishing as a prominent world center in agriculture, military technology, and salt water reclamation, and as well as classical music and the graphic arts.

To avoid discrimination as well as to seek assimilation, American Jews changed their last names. Cohen became Kahn and Cahn. Greenbaum became Greene. Goldstein became Golden. Shapiro became Shepard. Entertainers could be partially excused for the sake of the marquee, Cyd Charisse adopting that name from her given Tula Finklea, Tony Curtis having been born Bernard Schwartz, Kirk Douglas having been born Issur Danielovitch, and Jack Benny having been born Benny Kubelsky, but it didn't hurt bookings if they had what they referred to as "goyish," that is, gentile, sounding names.

Learning the lesson, Jewish parents would forego the traditional old-country Jewish first names. They would try to avoid the Irving's, Abe's, Sam's, Ben's, Nathan's, Max's, Bernard's, Myron's, Meyer's, Hymen's, Saul's, Bertha's, Tillie's, Libby's, and Rose's of their fathers and mothers and name their children Elyse, Iris, Renee, Brad, and Todd. Because Jews name their children after the Jewish name of deceased loved ones, this necessitated some imaginative phonemic manipulation. Parents would choose an English name of "Larry" or "Leslie" to honor and memorialize "Eliezer." Similarly, "Moshe" would be remembered as "Morris" or "Mark," "Chanah" as "Anne," "Miriam" as "Marla," "Binyomin" as "Bruce," and "Pinchus" as "Paul." Emmanuel Kimmel preferred to be called Manny. As I said, he named his business "Kinney," an Irish surname.

In the case of our family, both Vivian's father, Abe Sinetar, a businessman, and her mother, Adele Steiner, were Jewish. Because Jews ascribe the religion of the mother to the children, that connection being documentable, all of us Thorp kids are Jewish, and we were raised Jewish. My maternal grandparents' names "sound" Jewish, but "Vivian" was a good old American-sounding name.

Whether or not his associates knew his ethnic background, Manny's successful resume included being a bookie in his younger days and running the numbers game and other books in New Jersey. He was knowledgeable, connected, and passionate about his "career." This passion even led him to own several racehorses. By the time he met Oakley he had never served a day of time, and he never would.

Oakley paused, thunderstruck, to borrow Robinson Crusoe's description at seeing Friday's footprints in the sand, at this unexpected diatribe from what he was hoping would be his bankrolling bookie buddy.

"It's me trying it or nothing, huh?" he asked Manny. He turned to Julian. "Spring break, Julian?"

Julian nodded his head affirmatively, then checked in with Nancy. "Nancy?"

Vivian chose this moment to return with a candy bar.

"The Three Musketeers," Nancy said, referring either to the candy bar or to the trio of Oakley, Julian, and Manny.

"I can't gamble," Oakley reconsidered.

"You never played blackjack?" said an incredulous Manny.

"In grad school, we went up to Tahoe once, but not with these kind of stakes."

"What's the diff?" said Manny. "It's the same game. You want to try out your system, then we give you the cash and you do the playing." He stared at Oakley, who, by his pursed eyebrows, showed he wasn't convinced.

"Hey, Thorp," yelled Manny. He slammed his fist on the table, causing the coffee and tea cups and their saucers to jiggle. "You want to prove it works, then you play," his voice was suddenly raspy, showing that years of screaming had taken its toll on his vocal cords. "I didn't come up here to take no crap."

*　*　*

Other than the box of chocolates sitting on the kitchen table, all artifacts from the cordial business meeting had been cleared from the dining room table and either had been put into their respective drawers or the refrigerator or were being so placed. Vivian was placing the remaining silverware into the kitchen drawer next to the sink, and Oakley was putting the coffee and tea cups and saucers onto a cabinet shelf.

"On the contrary, I thought he was very well-mannered and I thought they were delightful girls," said Vivian. Oakley had been apologizing.

"Oh, I thought . . ."

"Nice girls. Polite," recalled Vivian. "'Double, please.' I wonder what side of the family they're on. His mother's or his father's."

"Well . . .," said Oakley. He knew Vivian was playing with him, and he was just waiting for his first New England nor'easter.

"Very nice. That was very nice." She inhaled deeply, to be able to maximize the volume and velocity of the cyclone.

"I am not a cocktail waitress for street-walkers!"

"Vivian . . .," said Oakley, not quite yet in begging mode. He was grateful that the cups and saucers had been safely placed in the cabinet. He was certain that they would have been swept up and carried off to Auntie Em's farm in Kansas.

Vivian reached into the box of chocolates and pulled out Madeleine's half-eaten piece.

"Here, you want a piece of chocolate? It's already been taste-tested."

"I'll admit that . . ." Vivian wasn't ready for excuses. Making up was still miles away.

"Will the real Edward Thorp please stand up?" she said. "Is it Edward Thorp number one, Edward Thorp number two, or Edward Thorp number three? The jazz trumpet player and mathematics professor? Or the gambler who consorts with prostitutes and gangsters?"

"What's number three?" Oakley asked. He hoped, with only a small probability of success, that a smart-aleck joke could dissipate the storm.

"You're not going to charm me, Edward Thorp," she said, rejecting the attempt. "Do you know the secret of being charming? It's to make others think they're charming. You're not doing a very good job right now."

"It's a business deal," Oakley explained. "At school you have to consort with the government to get grants to support your research. Here, it's a different sort of funding." Maybe reason would help.

"Funding? Funding? I have to bring these kinds of women into my home and be their cocktail waitress to get funding?"

"It's just a temporary association. He's our road to fortune, Vivian. He's got street smarts, and money. This is not about money, but we get 10%. and if I double his money our $1,000 is nearly two months of my salary."

"Ten percent? My father should teach you something about business."

"And look, Vivian, he was, uh, persuasive."

"I thought I knew you. I was searching for someone like you, searching like a river that can't find the sea, and I thought I found you, but now . . . Why do you have to do this anyway?" Vivian had thrown in a lyric from "Gingi." Perhaps her anger had short-circuited her mind and she had run out of words of her own.

"The computer simulations you and Julian did. They show your system works. Isn't that enough?" she begged.

"It's not enough. I've got to prove to these skeptics that I'm not just in an ivory tower, damn it," Oakley said. He had given up trying to appease Vivian. The storm now burst from deep within him. "They were mocking me at the meeting. My father, I love my father, but he always would say pure math is for people who walk on clouds. He worked as a bank guard, and the family moved to L.A. so that he and my mother could work in a defense plant. He believed fervently in the power of education, but he didn't want me to struggle financially. He would introduce me to his friends, 'This is my son, Edward. He wants to be like the King of Laputa. He thinks he can make a living walking on clouds.' They'd look at me just like they did at the math conference and laugh and mock me."

"Oh, Edward." She had never heard this chapter of the Thorp family history before.

Oakley turned his back to Vivian, stared out the kitchen window, and then turned back to her.

"And there's another thing. I'm a scholar. Most of us aren't very pretty. We don't have constantly beaming smiles or great personalities. We've got brains. Our product is the product of our brains. I mean look at me. I wasn't the most athletic or handsome or toughest kid in the neighborhood growing up in Chicago and I'm not now."

"I think you're very . . ."

"I skipped grades and found myself in classes with older, bigger kids. To survive in the neighborhood I had to be a wise guy, use my brains. Later I fell in love with numbers and I was good at it. I'm still good at it. They're my life, and I'm proud of my work and no one, not the reporters, not my father, not the editorial writer at the *Washington Post*, no one is going to mock me and my work, especially when I know it's right. You remember what that pit boss at Barney's said? 'When a lamb goes to the slaughter, the lamb might kill the butcher. But we always bet on the butcher.' Well, I have to prove that he's wrong.

"This work on probability is like a capsule of the great scientific projects. It may rank with that of Cardano, Pascal, Laplace, and Poisson, and their works live on forever. I'm not going to be denied that chance. And, even if it doesn't gain that stature, no one is going to deny me the thrill that comes from learning and proving things nobody else in the world had ever known, especially when it's right."

He paused and reflected. "This isn't exactly the image my father had for me."

"Oh, Edward. I didn't know. But he loves you." Vivian was feeling the kind of grief you can only feel when you've inadvertently hurt someone you love.

"Yeah, maybe. But it's true what he said. Mathematicians prove lemmas, conjectures, theorems. You know how you get a Ph. D in math? You do a proof," Oakley said. Now he was mocking himself, thought Vivian.

"Pages and pages of lemmas and derivations. That's it. No real use. I have to do a different sort of proof, Vivian. First, I'm going to shut those people up. Then, I have to prove to everyone, including myself, that I can do more than walk on clouds."

Part II

Field Trips

13

The Liechtensteiner Polka

Vivian must have said "okay."

Manny had secured lodging for all of them, Vivian and Oakley, Allan and Joan, Julian and Nancy, as well as an enormous luxury suite for himself and his nieces, at the most prestigious gambling location in Reno, the Mapes Hotel. The first high-rise hotel/casino complex in the entire state of Nevada, located on the northeast corner of Virginia Street and the Truckee River, it was the destination of both celebrity visitors and performing acts booked in the top floor The Sky Room. Marilyn Monroe, Clark Gable, Frank Sinatra, Milton Berle, Jimmy Durante, Sammy Davis, Jr., and many others had either stayed or performed there, hanging out during leisure hours at the ground floor Lamplighter Bar. Vivian and Oakley's act would never have been booked at the Mapes Hotel.

It was just as well. To keep business and pleasure separate, Manny and Oakley agreed that they should not stay where they would gamble. They would gamble at Harold's Club.

The sign above the stage proclaimed the name of its more modest entertainment venue to be the "Pair-o'-Dice Lounge." Below that was a smaller sign reminding the potentially inebriated patron that the lounge was in Harold's Club.

Several cocktail tables sat in front of the stage, and more filled the rest of the 80-person occupancy lounge. A modest-sized dance floor faced the right side of the stage as viewed from the audience. The stage footlights were turned off, but the house lights were on, providing sufficient lighting for the visitors.

Julian and Nancy sat at one cocktail table next to the stage. They had been a couple for almost 6 years and, although Julian was now 31 years old, marriage was not an issue, as much as their parents would have *kvelled* over it. Traveling to Reno together was only a quantum commitment greater than traveling

© Springer International Publishing AG 2017
L.M. Golden, *Never Split Tens!*, DOI 10.1007/978-3-319-63486-9_13

together to Cape Cod. Manny and his alleged nieces, Madeleine and Renee, who had accompanied him to Reno, sat at an adjacent cocktail table. They were drinking, no doubt a vitamin-packed fruit drink of sort for Madeleine.

Allan, Joan, and the lounge manager, Bernie Kurtzman, in his 60s, sat at a third table in front of the stage. Two other chairs were at the table, abandoned by Vivian and Oakley for their trip to the stage. Chairs were sitting, upside-down, on the other, unoccupied cocktail tables. Allan wore a coat and tie. The tie was sloppily knotted, a giveaway that coat and tie were not Allan's habitual attire. Allan had finished his post-doc at UCLA, and he and Joan had returned to Berkeley, where he had obtained a teaching position. Academic life did not require a coat and tie.

If you challenged the speed limit, it was a short 3¼ hour drive from Berkeley to Reno through Donner Pass, that of the ill-fated Donner party pioneers, and over the Sierra Nevada. This was a scenic California trip. Highway 80 passes through Sacramento to the Donner area and, having negotiated Donner Pass, a ride over the highest point along Highway 80 of the Sierras whose treacherously steep descent to the east requires tire chains and cautiously low speeds during the winter, you drive a few more miles along the north shore of Donner Lake into Truckee and then follow a serpentine path through the Truckee River mini-valley into Reno. You frequently drive next to the river itself, crossing it back and forth no less than eight exhilarating times before reaching the gambling mecca.

Actually, to tell the truth, the historical Donner Pass lies about a mile south of Highway 80, but everyone refers to the route as going through Donner Pass anyway. It makes it sound like you're more of a brave old West pioneer and you gain the sympathy of those who have had to install chains.

Allan had learned well that after a winter storm, the residents set up shop at the top of the pass selling chains and chain-installation services. You have to decide whether to spend the $10 for installation alone or freeze your hands putting on chains yourself. The residents of the Donner area must either be proud of its name or aware that it may attract historically literate tourists who may or may not be considering cannibalism. It contains a Donner Avenue, Donner Lake Road, Donner Pass Road, Donner Creek, Donner Lake, and Donner Memorial State Park, in addition to numerous cafes and other businesses with the name. The traveler obtains some designation relief reaching Truckee a few miles to the east.

Allan and Joan watched as Vivian, Oakley, and a jazz trio composed of a pianist, bassist, and drummer stood or sat on the stage, as befitted their instruments. They were about to begin performing, without microphones. Bernie alternated looking at the stage and at a photograph and resume of the jazz duo.

Vivian spoke the first line of "On a Slow Boat to China," a nice artistic effect, and then sang as Oakley accompanied her with jazz licks on his trumpet.

"All I know is," she said.

The trio began to play as Vivian started to sing.

I'd love to get you on a slow boat to China,
All to myself, alone.
Get you and keep you in my arms evermore.
Leave all your lovers weeping on the far away shore.
Out on the briny with a moon big and shiny
Melting your heart of stone.
I'd love to get you on a slow boat to China,
All by myself alone.

Vivian walked over to the piano as Oakley began to play a jazz chorus on his trumpet. The trio was happily playing behind them. They had realized that this wasn't the usual Pair-o'-Dice Lounge type of offering. As Oakley came to the final bars of his solo, Vivian walked back to the center of the stage. She continued the song,

Out on the briny with a moon big and shiny
Melting your heart of stone.
I'd love to get you on a slow boat to China,
All to myself a -
All to myself a -
All to myself alone.

It's a great tune, and Vivian and Oakley performed it professionally. The house trio looked at each other, smiling. This was a duo with whom they'd like to, as they say, go on the road.

Manny and his nieces, Allan and Joan, Julian and Nancy, and Bernie applauded. Allan put his forefinger and middle finger into his mouth and whistled quite loudly, especially for a mathematician. It really didn't have to be so loud in the nearly empty Pair-o'-Dice Lounge, thought Nancy. Vivian and Oakley smiled, turned around to thank the members of the trio, then walked over to Bernie's table and sat down.

Bernie had liked the music. "Play a little more, guys," he said to the trio. They had been called in for the purposes of the audition, and, if Bernie had not been the guy who would hire them, they probably would have demurred.

The trio, in good judgment, allowed themselves to be manipulated, as most artists must, and played a ballad. They were feeling jazzy themselves, and chose an up-tempo "Blue Moon," with the drummer displaying acumen on the brushes.

"Can I have a dance with the prettiest girl at the prom?" said Allan to Joan.

"Shucks. Sure, cowboy," she replied.

They stood up and walked to the dance floor and started to dance.

Manny, too, felt the groove. Showing ultimate *chutzpah*, he turned to Nancy.

"I ain't much of a dancer, but it would be a great honor if you'd join me on the floor." Nancy thought it was probably inappropriate to dance with your nieces unless it's a wedding. Nonetheless, she looked at Julian for approval.

"Go ahead," he said. "I'm not a dancer at all."

"It's easy, Julian. You just move your arms and legs with a complete lack of coherence."

"Nah," he said. "Just enjoy yourself."

Nancy and Manny walked over to the dance floor and started to dance.

"First, let me tell ya'," Bernie said to Vivian and Oakley. "I'm very impressed with both of ya'. Very talented kids. Can I ask you a question?"

"Sure," said Oakley.

"You know," he began, "you gotta be twenty-one to play this lounge. Do you know that? Same thing with all the lounges in Reno. You kids look like teenagers, college students."

"We're just out of college," Vivian answered, telling a half fib about Oakley's credentials.

Oakley waved to Allan and Joan to come over and join them.

"Our manager should be in on this," he explained to Bernie. Indeed, Bernie Kurtzman had thought it was strange that the manager of the act had not been sitting at the table immediately after "On a Slow Boat to China" had ended. For their part, Vivian, Oakley, and Allan hadn't spent a lot of time rehearsing for the audition showdown in the Pair-o'-Dice Lounge.

Oakley gave Allan a half glare as he and Joan walked over and sat down at the table.

"Yeah," said Bernie, unaware of the drama. "See, that's the problem." He directed himself to the duo's manager. "These kids are just out of college. The people who come into this lounge here are in their fifties, sixties, seventies." He pointed over to Manny. "Like that guy over there with the blondes. They wouldn't relate."

Allan adopted the part of the manager. "The kids need a break, Bernie," he said. "And the act worked in Tahoe just a couple months ago."

"Sure it did," responded Bernie. He had heard the same line over 70,000 times in his career. He turned to Oakley and Vivian.

"Kids, you're great. Great. Really. But this is a "Liechtensteiner Polka," "Yellow Rose of Texas," "Peg O' My Heart" kind of crowd, ya' know? I mean, if someone asks for the "Liechtensteiner Polka," can you do it?"

"Uh, we don't really play polkas," said Vivian in a line they would relate, trust me, numerous times to their children yet to be born.

"See? Okay," said Bernie. "Look, I'll put your photograph and resume in the file." He looked at Allan.

"Allan, bring 'em back, two, three years." He turned at Vivian and Oakley and gave them some stern fatherly advice. "Learn a couple polkas. You know, da da dat dah, da da dat dah," singing the first few measures of the "Liechtensteiner Polka." "More tunes, then we'll try again. Right now, it's not a good match."

Vivian cried softly. Oakley must have been amazed at what a great actress she was. Mom was really, really into this.

"Come on," said an obviously touched Bernie. "It's okay. Look, ever spend any time in the casinos?"

"No," said Oakley.

"I keep my clients under control, Bernie. This is strictly a business trip," said Allan. Now they were all fibbing, dissembling, misrepresenting, uttering falsehoods, practicing deception, wallowing in the untruth, lying! They were all relishing the skillful theatrical ensemble they had created. It was, however, in the interests of science, Oakley thought to himself.

"That's good," noted Bernie.

He pulled a small piece of paper out of his shirt pocket. "Look, here's a toke. How many of your friends did you come up with?"

"One other couple," said Vivian, without, thankfully, pointing to their manager.

"Okay," said Bernie. This is a toke for twenty dollars for each of ya'." He scribbled some notes on the piece of paper. "B K, for Bernie Kurtzman," he said showing Vivian the note. "Take it to any of the pit bosses. They'll take care of ya'."

"Thank you," said Allan at Bernie's graciousness toward his clients.

"Play the slots, a little roulette, maybe a blackjack hand or two," Bernie recommended.

"Well, I know a little bit about roulette," Oakley volunteered. "My friend Claude showed me. But blackjack. Too much math for me. It scares me."

Vivian covered her mouth with her hand and rubbed her upper lip as if to scratch an itch. If she hadn't, Bernie would have detected a wide grin.

"Don't be silly," said Bernie. "Look, just enjoy yourselves. And don't get downhearted about the act. Try it again. You guys are good. Just not a match right now for the Pair-o'-Dice Lounge."

"Sorry for wasting your time, Bernie," said Allan, smiling. This gave him an excuse for nearly laughing out loud at Oakley's comment. "Like you say, not a good match."

"Okay, look. I was young once. Here are, let's see . . ." He pulled out additional slips of paper from his shirt pocket. "Six vouchers for dinner at the Eta On Rish, our classiest restaurant. Sort of Frenchy, Greeky kind of place. The alphabet soup. It's incredible." Vegetarian. No factory farm inhumane stuff.

"We really appreciate all of that. Thank you," said Vivian.

"And don't forget the saganaki, the flaming cheese," added Bernie.

"Oopah!" bellowed Allan.

Everyone who knew Bernie Kurtzman knew him to be a gracious, compassionate, kind person. Our actors certainly had discovered that. He had moved to Los Angeles from Chicago at age seven with his parents Sam and Pauline (Berman), who changed her name from Pearl, and younger brother Myron in 1926. His father had seven siblings, Jack, Abe, Phil, Ben, Sara, Faye, whose name had been changed from Fanny, and Ann. They and their families had all lived in an apartment building in Chicago owned by their father, Bernie and Myron's grandfather. In the economic decline

OOPAH!

leading to the Great Depression, the Kurtzmans lost the building, and the entire clan, one by one, had relocated to Los Angeles.

Pauline had two siblings. In a story representative of Jewish migration to the United States from Eastern Europe around the turn of the century, older sister Gussie had moved to Chicago. Pauline and her brother Oscar, being younger, had remained in Eastern Europe, perhaps Poland, with their parents, but they immigrated to Chicago at a young age after their mother died and their father remarried, to a woman with a sizable number of children. Their stepmother was not particularly pleasant to her stepchildren Pauline and Oscar. In Chicago, they were reunited with Gussie. Oscar would also move with the Kurtzmans to Los Angeles. Their move was accompanied with mixed feelings, Pauline and Oscar leaving their sister Gussie, and adopted Chicago in spite of the opportunities and better weather of southern California. Many Jews were moving to the west coast for the better weather. The various Kurtzman brothers would go into business together and eventually thrive in Oriental importing and mens' wear businesses. Oscar would work for two of them.

Bernie lived in a modest home in Los Angeles off of Robertson Blvd. He was a successful garment salesman at an upscale Los Angeles men's store, and his brother became a dentist and lived in Sherman Oaks. Both married Los Angeles-raised Jewish women. Uncle Phil, as he was known, remained close to the families of Bernie and Myron and worked for Universal Studios in the wardrobe department. He eventually married a woman he had been "dating" when he was in his 70s.

When he was in his 50s, Bernie's two daughters and his son now grown, Bernie and his wife Freda decided to move to Nevada. It was going to be just a temporary change of life. Las Vegas was too frenetic for their gentle, homey personalities, *haymisha*, in Yiddish, so they decided on the Reno area. Although he had planned on an early retirement, he got a little antsy and fell into the lounge manager position at Harold's. He enjoyed the job, particularly helping young talent advance their careers. He felt that Oakley and Vivian, at this point, needed further preparation.

During this business meeting, Manny and Nancy were dancing. He wasn't so shy as not to try to pull Nancy close to him. Nancy wasn't so weak physically as not to be able to keep him at a respectable distance.

"See, Madeleine and Renee, they ain't my kind of girls. *Trafe*," he said, using the Yiddish word for non-Kosher food. This was his first admission of the obvious, that Madeleine and Renee were not his relatives.

"Oh," said Nancy.

"They got no class, ya' know?"

"I'm sure they're very nice," Nancy said.

"No, my mom, *a la sholom*, she always said I should go with a girl like you. You know, intelligent, nice Jewish girl."

"Oh. I'm flattered," said Nancy. As much as she sensed the come-on, she had to be polite. This was a business trip for Julian and his friend Dr. Thorp.

Manny laughed at Nancy's reply. "I'm glad. You know, my mom, she used to tell the *goyem* at Christmas 'Merry *Kratz Mer*.' Scratch me."

Nancy laughed. That was clever. They danced a bit.

"Do you know what *bei mir bistu shein* means?" he asked.

"Isn't it a song?" responded Nancy, politely but again tersely.

"Yeah, by The Andrews Sisters. It means 'to me you're beautiful.,' ya' know basically a grand Jewish girl. I used to try to date *shiksies*. I'd ask them, 'would you consider a thing with a short, balding, Jewish millionaire?' They had a *goyishe kop* and they looked puzzled, so I said, 'Well, how about three out of four?' "

Nancy laughed, but again didn't respond.

"Are you religious?" said Manny, trying to revitalize the conversation.

"I guess," she said. She would remain polite, but wouldn't provide any hope to Manny by offering information about herself.

"Not my family. On *Pesach* we ordered in ribs and shrimp."

Nancy laughed. That was a good joke, although it sounded as if it had been lifted from some, well, lounge act.

"Yeah. All Jewish holidays, ya' know, are the same. They tried to kill us, we survived, let's eat."

Nancy laughed, even though this joke also sounded lifted. Manny got a little serious now, going in, he undoubtedly hoped, for the kill. He tried to pull Nancy closer toward him.

"You know, older men, they know how to treat a girl right."

"Oh? Is that so?" she replied, resisting.

"Yeah, it is so. Especially guys like me who gots lots of money."

Nancy felt it was time to politely let Manny know that his attempts would not succeed.

"Mr. Kimmel, I love Julian. I want to grow old with him."

"Hey, wit' me you're half way there," he persisted, although this time with an original line.

Nancy had to laugh. It was, she had to admit, an admirably quick comeback.

"And what am I anyway, pickled herring?"

"Uh, Mr. Kimmel, I think the expression is 'chopped liver.'"

"Chopped liver, pickled herring, gefilte fish. What's the difference?"

Nancy laughed politely, although she felt manipulated, having been obviously suckered into a response. Manny stared at her and laughed, hoping he was making some progress. He showed her his diamond ring.

"That's nice," she replied, returning to the terse manner with which she had participated in the entire conversation.

"Yeah. You know I'm a self-made man. Born dirt poor. Dirt poor. If I hadn't been born a boy I woulda had nothin' to play with."

Nancy laughed, but not as genuinely as she had to the growing old line. This one also seemed to be lifted from someone's lounge act.

"And with me, you don't gotta worry about kinky stuff. Not me, but I knew a guy who became a transvestite. He wanted to eat, drink, and be Mary."

Not bad, Nancy thought to herself, even though it was a pun.

Reading Nancy's reaction, Manny decided to make his move.

"Yeah. You know, maybe, later, you know. Maybe," said Manny, displaying the pessimism that Nancy's reactions to him warranted. "Later. You, you wanna come and visit me in my room, you know?"

"Oh, I'll do that," said Nancy.

Manny was happily surprised, even shocked. "Yeah?"

"Yeah," she replied, dropping her arms from Manny's shoulders. She walked away and turned for a second toward him.

"As soon as the Dalai Lama starts studying for his Bar Mitzvah."

14

Stargazing

The afternoon following their pseudo audition, Oakley was ready to set the foundation for trying out the complete point count system that he and Julian had developed. First, as they had done at Barney's in South Lake Tahoe, they would put the casino off guard by playing like unsophisticated young kids, making stupid moves and betting low stakes. Here, though, they would focus on using irrational so-called systems.

The four, Allan, Joan, Vivian, and Oakley, sat at a blackjack table at Harold's. The other seats were vacant. It was an understated place, with muted green and gray colors. Oakley would get to like Harold's Club both because of the quiet atmosphere and because he would always seem to win there.

Each player had bet a single $1 chip, placed neatly in the very center of their betting squares. Cocktail glasses and glasses of water stood in front of each position. The cocktail glasses were full. Even the water was barely touched. These poor young neophytes were so concerned about losing their weekly lunch money, they weren't even drinking a glass of water.

Two cards, a six of Diamonds and a five of Clubs, lay face up on the table in front of the first seat to the left of the dealer, Allan's place. Two cards lay on the table in front of each of the other three occupied seats, all face down. Two cards, one face up and one face down, lay in front of the dealer. The up card was a seven of Hearts.

Jackie, a female dealer, about 25, stared at Allan. Clive, the pit boss, stood next to Jackie, also observing Allan.

© Springer International Publishing AG 2017
L.M. Golden, *Never Split Tens!*, DOI 10.1007/978-3-319-63486-9_14

Allan's eyes were closed as he opened a Bible to a random page. He stuck his forefinger onto the open page of the Bible, then opened his eyes and looked at the Bible. He uttered a sigh and showed the configuration of the Bible pages and his finger to Jackie.

"Hebrew words," he said. "I stand."

"Religious, huh?" concluded Jackie.

"I believe that the Lord watches over me and guides me," Allan said righteously. "If my finger points at an English word, I take a card. If it's a Hebrew word, I stand."

"But you've got eleven," protested Jackie.

"The Bible says stand," said our sanctimonious one.

"Oh, God," said Jackie in understandable disbelief.

Allan glared at her. This was a firm believer in the ways of the Lord, and any heretical behavior would, if he had any means of communicating with the Supreme Being, be severely punished, if not in this life, then in the afterlife. Allan was proving to be a good actor.

"Sorry," apologized Jackie, not wanting to irritate a client. She cleared her throat. "That's quite a system."

"It's the best," Allan guaranteed her.

He communicated his conviction with a mixture of near sexual passion and the deepest religious faith. Poor guy, Jackie must have thought. His girlfriend probably wore an overcoat to bed.

"It's better than the other systems we see," interjected Clive. "Sure, sometimes they get lucky. Just like when a lamb goes to the slaughter, the lamb might kill the butcher. But I'll always bet on the butcher."

Oakley, remembering the butcher story from Barney's, thought it must have been a line taught at the Pit Boss Institute of Technology.

"We're vegetarians," chimed in Vivian, she too remembering the episode at Barney's.

"You meet all kinds," said Jackie, trying to lighten up the table.

"There was a guy from Illinois," recalled Clive. "Called himself the Stargazer. Had some kind of magic toy slot machine. He rolled it before every hand. If it came up cherries, he bet. If not, he laid out. And darned it but he usually seemed to bet right."

"Heresy!" exclaimed Allan. "I put my trust in the Lord, not in pagan instruments or the devil's toys." Oakley thought that Allan was putting it on a bit too much.

"A real character," continued Clive. "He used to order drinks, liqueurs, the expensive stuff, and never drink 'em. Had a bottle hidden behind one of the slot machines. He'd leave the table and pour the booze into the bottle."

"Talk about coming for the free drinks," said Oakley.

Clive pointed to the cocktail glasses on the table.

"Sort of like you kids, huh? It's fine. But maybe you should learn how to play blackjack a little better."

"This is exactly how I always play," said Oakley. "Same strategy. I stand at twelve or above so I never go over twenty-one."

"Me, too," added Allan. "I always count on the Lord."

Oakley would later tell me about the Stargazer. Apparently he had a lot in common with him, and they by some quirk of fate, became acquainted through mutual friends in the UCLA math department. The Stargazer was a graduate student in astronomy from Berkeley, although Vivian said she had never met him. He apparently had been one of the founders of the University of California Jazz Ensembles, the group for which Vivian was a vocalist. You recall she met Oakley at the Pacific Coast Collegiate Jazz.

He had played very low stakes poker games with his twin brother and buddies during high school, and a little blackjack. In preparation for the first trip of the band to the Reno Jazz Festival, under the direction of David Tucker, who was also the director under whom Vivian sang, he read a book on blackjack by Oswald Jacoby, the bridge maven that he had bought at the famous Moe's Bookstore on Telegraph Avenue. To prepare, he and his girlfriend, Diane Dring, whose mother had been a vocalist herself, practiced in the din of the Bear's Lair snack room on the Berkeley campus. Having never been to an actual casino, he wanted to simulate its noise and distractions.

The Stargazer was sufficiently successful on that Reno Jazz Festival trip that he would make monthly trips to Reno for the rest of his graduate student days at Berkeley. He'd leave on a Sunday afternoon, when the other casino players would be returning to the Bay Area, and stay through Tuesday afternoon. The chartered bus that the residents of Chinatown would take to Reno and back, the "chicken soup express," so named because the Chinese would bring containers of chicken soup with them from San Francisco for the 3 1/4 hour trip, would be packing up to leave as the Stargazer arrived. He always stayed at a small hotel, the El Cortez, that in any other city would have been a brothel or vagrant's destination. It's still there in Reno, west of Virginia Street, a small hotel, with a red neon sign, and single elevator right there in the small, do-not-tell, front lobby.

Despite betting low amounts, the Stargazer's frequent visits began to draw attention. After all, even if you win only $50 at a shift, that's $100 that the casino has to make up to earn their $50. So, as Oakley told me, the Stargazer came up with booze as a camouflage technique.

If you're a serious player, you should never drink while playing. Among other things, it interferes with your rational decision making. The Stargazer, who Oakley told me was also an accomplished actor, played the lush to hide being a systems player. Before sitting at a table, he would dab some booze on his neck, face, and hands so he would emit that telltale fragrance. This, Oakley recalled, was exactly one of the camouflage techniques suggested by Manny during the Brookline visit.

Immediately upon sitting down, the Stargazer would order a drink, straight, no rocks, and if it didn't arrive quickly he'd make a small fuss. He didn't order the usual Scotch or bourbon, but the expensive liqueurs, amaretto, kahlua, and crème de menthe. He'd pretend to take a hit from time to time but wouldn't really drink.

When his session was over, he would stumble slightly on leaving the table saying, "Time to try the slots." Even that statement was a camouflage, implying he was a gambler. No system exists at the slot machines. He would walk over to the rows of slot machines and pour whatever drink he had into a bottle he had hidden between a couple of the slot machines. Judging from Clive the pit boss' comments, the casinos eventually caught on to the hidden bottle trick. Nonetheless, as you'll see, Oakley would learn from the Stargazer. He would later develop and use camouflage techniques of his own.

Here, at Harold's Club, they were already using some. They were pretending to be mooching drinks, again as Manny had suggested, and playing based on faith and superstition, in Allan's case, and playing using naïve systems, the "do not bust" strategy, in Oakley's case.

*   *   *

It was evening. Oakley, Vivian, Allan, Joan, Julian, and Nancy had enjoyed a nice dinner at the Riverside Hotel, later to be known as Jessie Beck's Riverside. To get there from the main casino strip on Virginia Street you had to cross what everyone referred to as the river. It was, by name, the Truckee River, although in Reno it was no more than a large, lazy stream rather than a broad expanse reminiscent of the great American rivers or even the rushing Truckee further up the mountains near its source.

They sat down on a couple of benches facing the river and listened to the sounds of the reclaimed desert night. Oakley had been thinking about the entire escapade, learning a system, trying out the system, proving the skeptics wrong. He was wondering about the place of such ventures in the fabric of one's life.

He had pointed out Venus in the sky, only a month past eastern elongation, making it startlingly bright in the early evening hours, and his friends were enthralled by it and the multitude of other stars visible in the non-polluted skies over Reno. One day, he had told Vivian, he wanted to build an astronomical observatory to indulge his interest in that subject. Some birds scampered along the river edge, looking for seeds fallen from the trees growing on the river bank. Joan noted that they were probably vegetarians and that the worms there were safe. They watched the birds.

"We're born to die," Oakley said, breaking the silence. "You start to die when you watch those you love die. That's what 'over the hill' means. What hill? You're over the hill when those you love have died."

Edwin Hubble

ASTRONOMERS STAY UP ALL NIGHT

"Whoa," exclaimed Allan. "Where did that come from, Ed?"

"That sounds like a Jewish piece of wisdom," noted Julian. "We're born to die."

"You're exactly right, Julian," said Vivian. "I learned the phrase from my parents, Abe and Adele. And this guy learned it from me."

"I don't care if it's your dog or your parakeet or your parents," continued Oakley. "That's when the sense of humor starts to go and you stop caring about life. You let yourself go physically and don't eat well. You're getting ready to die."

"Well," said Vivian, "I hope most people wait a few decades after those events have hit their lives."

"All of life is full of highs and lows," said Nancy. "My uncle Eliezer Moshe always says, 'As night follows day, *tsouris* follows *nachas*,' misery and joy."

"So why do we screw?" asked Allan.

"Where'd that come from?" Vivian said, repeating Allan's comment.

"That's not the question," said Oakley. "That's the wrong question. The question is why does it feel so good to screw?"

"I remember reading *Of Human Bondage* by Somerset Maugham in high school. It tried to answer the question of 'what is the meaning of life?'" Julian said, the beginning of a presentation that Nancy had heard before. Nancy thought that he must have been seriously affected by that book.

"If you want to learn about the meaning of life you've got to look at sex," Allan persisted.

"Now we're talking," said Joan, playfully pushing her elbow into Allan's ribs.

"Porno movies!" said Allan. "You can learn everything you need to know about life from porno movies."

"You are so silly, Allan," said Nancy.

"I place great import on them because I'm so well endowed."

"Don't get too personal, Allan," cautioned Joan.

"I'll just say that I'm three inches," he countered.

"I don't think most statisticians would classify that as 'well endowed,'" observed Oakley.

"Oh, I'm sorry. That's three inches . . . from the ground," replied Allan with surprisingly adroit comic timing.

They all laughed heartily.

"Isn't the meaning of life simply to provide food for the species next higher on the food chain?" suggested Vivian.

"If you're a carnivore," replied Allan.

"No," answered Oakley. "Plants survive in part on soil, which is in part decayed vegetable matter. Herbivores, be they plant or animal, survive as being higher on the food chain. My wife is undoubtedly right, as usual."

Vivian snickered. "Everybody look at Vivian," commanded Oakley. "Notice the wrinkled nose? She wrinkles her nose when she laughs."

"I used to think, and I thought this was a great answer, that the meaning of life was to gain immortality," said Julian, getting back to the moribund.

"Oh, I like that," said Vivian. Her husband's colleague Julian was a philosopher.

"We live so that we never die," Julian repeated the philosophy of death he had described to Nancy the first time they had visited Buckingham Fountain. "Some people give money to name a library after themselves. Some write books. Some get rich enough to endow an academic chair in their name. Maybe one of us, God willing. These all accomplish their end of achieving immortality."

"That may be true, and it may not be true," said Allan, perhaps playing the party pooper as well as the devil's advocate. "But, have any of you ever watched two flies having sex?"

"Not recently," admitted Oakley.

"Or two moths? Or two spiders?" Allan asked, obviously being serious about the line of questioning.

"Moths, no. Spiders, yes," said Oakley, supporting his friend Allan Wilson in whatever monologue he was rehearsing.

"They stay attached forever," said Allan, "as long as you have patience to watch. That means, my friends, they experience the same pleasure we do."

"Or they didn't read the complete instruction manual," Joan noted rakishly.

"Mammals, fish, insects, reptiles," Allan continued. "Amphibians, whatever they are, for that matter. It's ingrained in evolution."

"And they're not thinking of immortality," noted Julian. "Perhaps I should reconsider my hypothesis."

"Well, at least in humans you can keep the immortality concept," said Nancy, still believing in Julian's hypothesis, and wanting to buoy up her man.

"Yeah," said Oakley. "But love of sex came with evolution, from four and one-half billion years ago. What's the drive? Are you saying the meaning of life is to enjoy sex?" Oakley asked Allan. "Then you do get pleasure and propagation. And it's the same in every animal."

"Maybe even in plants," threw in Vivian.

"Pleasure and propagation," added Julian, perhaps conceding. "Those are worthy evolutionary goals."

"That's an interesting idea, guys," said Allan.

"Where does being a mathematician come in?" asked Oakley.

"And gambling?" added Julian to the conundrum.

"We'll discuss those issues next time, class," said Professor Thorp at the tightening intellectual bottleneck.

"No, really, what's the point of intellect in general if the meaning of life is pleasure and propagation?" asked Nancy. Maybe she was still hoping for votes for Julian, or maybe she had just decided to join in the direction of the discussion.

"Obvious. It gives pleasure," answered Oakley.

"Okay, I'll go along with that," said Allan.

"But, if there are alternative means of gaining pleasure, then wouldn't they provide a disincentive to sex?" suggested Julian. "Then intellect is counterproductive to your suggested meaning of life."

"I don't think so," countered Oakley. "It's like the uncertainty principle. Sex and intellect are conjugate pairs."

"You mean conjugal pairs, darling," said Vivian.

"No, conjugate in the uncertainty principle sense."

"Right," agreed Allan. Vivian was only slightly embarrassed at her mathematical *faux pas,* grateful that her lack of knowledge of mathematical physics hadn't led Thorp, Allan, or Julian to utter even mild words of scorn.

"You give a little in one area to gain a little in the other," expostulated Oakley, now firmly ensconced in professor mode. "Non-sex pleasure is still pleasure, but it's less sex. It means less sex."

"Right," said Julian. "Studying time eliminates some dating time."

"Let delta x be the contribution of sex to pleasure and delta p be the contribution of intellect to pleasure," said Allan, scribbling figures on a blackboard in the air. "You can't have refined, that is, precise, values for both," continued Allan, now leaving the girls behind, as well as perhaps Oakley and Julian, if not in the physics then perhaps in the reasoning.

"And immortality fits right in, Julian," Oakley reassured him, puzzled a bit himself at Allan's application of Herr Professor Heisenberg's uncertainty principle.

"Well, you can't get laid in the grave," said Allan. "So it must give you pleasure."

"Which?" said a puzzled Nancy.

"Uh, although you can't get laid in the grave, your intellectual contributions continue, providing immortality and pleasure to your, well, ghost, I guess, and others." Allan had gotten a little far afield even for himself.

"What a great name for a country tune," stated Joan. "You Can't Get Laid in the Grave."

"I'll forget you said that," Oakley said to her.

"Edward hates country music," Vivian said.

"They all sound like they have terminal nasal congestion," Oakley said. "Give me Ella, Mel Torme, Frank D'Rone, or Chet Baker, or give me death," he exclaimed. "But, Julian's right," he added, getting back to the subject at hand. "Getting your name on a book or funding a library gives you and others pleasure."

"But probably not you if you're in the grave," noted Nancy, questioning Allan's thesis.

"I'm glad you didn't want to be a lawyer," Vivian said to Oakley. Allan and Joan joined Oakley and Vivian in chuckling at the early days of the Thorps' courtship.

"Right. What pleasure do they provide?" asked Julian, Socratically.

"I don't think they provide any," said Allan. "But they make money and that means more sex for them and their clients, if they won."

"And misery for those who lost," added Joan.

"It's the uncertainty principle heavily weighted on sex," said Oakley.

"Right," agreed Allan, his mind perhaps a little clearer.

"Most people have the opportunity to get pleasure from both sex and intellect," said Vivian. She elbowed Oakley in the ribs.

"Except Catholic priests," said Allan.

"That was one of the most ignorant restrictions in human history, and ironically it was made by the heads of Gregor Mendel's faith," said Julian.

"A case in point," Nancy noted. "He was after all a monk." Julian turned and stared at her with a querulous, mildly disdainful, look. Nancy really didn't have to state that to this highly educated bunch.

"If he had been alive then he would have explained its idiocy to the cardinals and the pope. Books were only available in monasteries," explained Julian.

"Yeah, you want to read books, you want to gain knowledge, then you can't screw anymore!" commanded Allan in his colorful manner, standing up. He stuck his arm into the air, stretched out his index finger, and preached. "And if you do you'll rot in hell for committing the sin!" Allan was practicing his evangelicalism. Joan grabbed him by the arm and yanked him down.

"If you wanted to be scholarly you had to suffer," lamented Oakley, somehow commiserating with scholarly brethren of long-ago centuries. "The prohibition selectively removed the high I.Q.'s from the Catholic gene pool for hundreds of years, from the 12th century until the printing press became common, dozens of generations," he observed.

"It just boggles the imagination. Couldn't the popes see that smart people have smart children?" wondered Vivian incredulously.

"The Catholics say you gain immortality by being a good person. That's another way," said Julian.

"I partially disagree," said Allan. "The Catholics say you gain immortality by being a good Catholic. It's the myth of the afterlife reborn as big business."

"I agree with Allan," said Joan. "The Catholics are hypocrites. You can disobey the Ten Commandments, you can bear false witness against your neighbor and sleep with your neighbor's wife and commit murder, but as long as you put five dollars in the collection plate and chomp on a cracker every Sunday you get to reside in heaven for eternity with Jesus and all the saints."

"Thank you, sweetheart. In other words," concluded Allan, "you can harm your fellow man, but it's all excused if you just donate. That's what their confession is, isn't it? Go into some dark chamber, tell me you're sorry, and then put some penance money in the plate and then Jesus and I will forgive your sin and it's okay if you got Mr. Whatzit convicted of a crime he didn't commit because you happily committed perjury out of grand revenge or you slept with Mrs. Whozat."

"One fall, I had to attend a conference at Cornell," said Julian. "So I took the Twentieth Century Limited train from Chicago to Syracuse and then went to the Greyhound station to take a bus to Ithaca. There, I got into a conversation with a young man from a small town in Iowa who was to begin his freshman year at Cornell. Somehow the conversation got around to Jews, maybe the large number of Jews who attend Cornell, and when I told him I was Jewish he looked puzzled and told me I was the first Jew he had ever met."

"There aren't a lot of Jewish corn farmers," noted Vivian.

"Yeah. So, he asked me if he could look at my forehead. I had no idea what he was thinking, but I said, 'okay,' and he started to part the hair over my forehead. After a couple seconds he said, 'You don't have horns.'"

"You're kidding," said Joan.

"No, really, he was looking for my horns. So he said, 'My pastor taught us that Jews have horns.'"

"It's part of the business model of the Christian church to propagate ignorance, intolerance, and bigotry," observed Oakley.

"Those intellectually-gifted Iowa preachers," commented Joan.

"Passing on their bigotry to their children and their flock," added Oakley.

"Well, what did you say?" asked Vivian.

"I told him Jews don't have horns," said Julian. He paused. "Cloven hooves, sure, but horns!?"

To this clever joke, uncharacteristic of Julian, they all laughed uproariously.

"The father of one of my girl friends growing up was a minister. She took me to church one Sunday and I'll never forget what he said," said Joan.

"What?" asked Vivian.

"He said, 'He who spills his seed upon the soil of the Earth shall reap the harvest of eternal damnation.'"

"How poetic," said Vivian, the English major.

"How visual," noted Allan.

"When I was a little boy in Chicago," recalled Oakley, "the Jews wanted to build a synagogue a couple miles from our house, right across the street from a Dominican priory. To get it approved by the town, they had to say the permit was for a bowling alley or an apartment building, but once it was obvious it was going to be a synagogue, maybe there was a sign posted, the good Dominican friars stood across the street on the sidewalk in their cloaks and yelled anti-Semitic taunts. Very nice people. Enlightened, progressive bigots."

It was Vivian's turn. "As far as I'm concerned, Catholicism and Christianity in general are untenable philosophically. The priests took perhaps the greatest intellectual achievement of all time, the concept of monotheism introduced by the Jews, and made a mockery of it. They started with the virgin birth mythology, then the trinity, and then the saints. No, I'm sorry, there is only one God, not three, not a hundred."

"Actually," added Oakley, "humans can have virgin conception, but since the female only carries the female gene, the child can only be a female. So Jesus, by this logic, had to have been a woman."

"That's interesting," chuckled Julian. "But the saints like Vivian said are my favorite, performing miracles. The patron saint of this and the patron saint of that. It's paganism all over again."

"Well," said Nancy, "at least the Catholics brought civilization with them to the New World."

"Right, those enlightened Spanish conquistadores, claiming the New World for Jesus," Allan observed. "They brought civilization, along with smallpox, typhus, yellow fever, bubonic plague, syphilis, influenza, physical atrocities, and murder, annihilating the sophisticated native American cultures, the Incas, Mayas, Aztecs, Toltecs, whatever, with their advanced agricultural, architectural, city planning, and astronomical knowledge. Sixty million by the end of the 16th century by some accounts. A significant fraction of the human race existing at that time. At least twenty million in Mexico alone. Some areas in the Americas had population loss of 90%, about three times greater than in Europe and Asia during the Black Death. Very civilized, Nancy."

"Quite a lecture," observed Joan.

"There are other theories of course," noted Oakley.

"Okay," conceded Allan. "But throw in the genocide and intolerance that was the Crusades and the Inquisition, mass murder of so-called pagans who would not adopt Catholicism and so-called heretics, the murderous Knights of the Order, Catholic-run extermination camps in World War II, and you can make the case that the Catholic Church has been the single most destructive force in the history of civilization, and over hundreds of years. Forget it if you were a Jew."

"Well, at least they gave Galileo a bed and Giordano Bruno was kept warm as toast," said Julian.

"Funny. I'm sure Galileo had all the water he needed. Unfortunately, that wasn't the case with Bruno," lamented Allan.

"Talking about drinks, my dance band the Deuces played for an Irish wedding," said Oakley.

"Okay," said Allan. "You have my permission to change the subject."

"Thank you, sir. So, two guys came up to the bandstand with a six-month old baby and gave him a sip of booze, and when he vomited they all laughed like it was a coming of age, getting drunk."

"The Irish. A nationality of alcoholics of low I.Q.'s," concluded Allan.

"It's an example of genetic drift," lectured Julian. "An event alters the gene pool of a population. It's why Ashkenazi Jews, who encouraged their rabbis and biblical scholars to have large families by supporting them financially, but were isolated socially from the rest of Europeans, have as a group high intelligence. Every Jew living today is probably descended from a rabbi or scholar." Julian's parents had schooled him well in the history of his people. "In contrast, the Irish lost much of their intelligence genes to the monasteries."

"That's why the highest aspiration of the Irish is to become cops and firemen," said Vivian, applying Julian's logic while attempting to alleviate the weight of the scholarly discourse.

"We need cops and firemen," said Nancy

."And politicians," added Vivian.

"We don't need them," said Joan.

"And they all have the same thing in common. Below average I.Q.'s resulting from selective depletion of the tail of the intelligence distribution," Oakley concluded. "It severely limits their career options."

"Some priests had children," said Joan, suggesting a flaw in the argument.

"I will not hear of such heresy!" remonstrated Allan.

"If I had to make the choice, I'd pick the monastery," said Oakley. It was his turn to poke Vivian in the ribs.

"Well, sweetheart," Vivian said to Oakley. "Then we just won't have children."

"I guess I'll just have to write a lot of books for my pleasure," teased Oakley.

"Books about our glorious love affair, I hope," she said.

"It's true," Allan interjected. "Every woman thinks the greatest story is about her love affairs."

Vivian, Nancy, and Joan laughed. "Well, I married Edward Oakley after being serenaded by his trumpet at a jazz festival. That certainly qualifies as one of the great love stories," said Vivian, poking Oakley in the ribs, again. He began to wonder if his spleen, whatever its function, would survive the evening.

"I knew a woman, before Joan, of course, who was writing a novel about her love life," said Allan. "And she repeatedly used the phrase 'all right' instead of 'okay.'"

"Nobody, and I speak authoritatively as an English major, uses the phrase 'all right' except unskilled novelists," Vivian agreed.

"And as a mathematician who married an English major, I concur," chipped in Oakley.

"And as a programmer who knows a mathematician who married an English major, I would also have to agree," said Julian, following the logic.

They all laughed at the joy their minds provided them.

"One of my writer pet peeves is the use of the word 'beautiful,'" said Vivian the English major. "It denotes nothing but approbation, that you like something. It's symptomatic of a lazy mind. Like using adverbs instead of searching for the right verb," she lamented.

"A classmate of mine at Cornell wrote a quantum mechanics text," Oakley added, "and the first sentence in the book was, 'Quantum mechanics is beautiful because it works.'"

"Exactly," said Vivian. "That only means that he likes it."

"He could have written that quantum mechanics is an ingenious construct of the human mind, demonstrating our ability to model nature in a manner that yields predictions and provides uncountable practical applications," said Julian.

"Right," said Vivian. "Then we know why he finds it attractive."

"When I was in high school, I took a summer class in public speaking," said Julian. "One of our assignments was to describe something beautiful without actually using the word."

"Now that was a good teacher," Joan noted.

"Yeah. Carroll Anderson," recalled Julian. "So I described driving with my parents through Florida en route to Miami Beach for winter vacation and watching the rows of orange trees stretch out in perfect alignment as far as you could see as you drove past them on the highway."

"Did you use the word 'beautiful'?" asked Nancy.

"Uh, I'm not sure. If I did, it was only once," said Julian defending any possible intellectual lapse, "until I knew that Mr. Anderson was making a black mark in his mental grade book."

"Beautiful," approved Vivian.

"And if I did, it was because I was staring at Diana Movius. She was an actress and singer in high school."

"Uh, who"? asked Nancy, playing at being jealous.

"Since we're talking about love lives, the concept of love-life novels was the basis of one of my great lines to women, B. M. J., of course," said Allan. He had skillfully used the reference to whomever that high school girl was to get back to the discussion which had been short circuited by reflections on the word "beautiful."

"B. M. J.?" asked Vivian.

"Before my Joan," said Allan.

They all laughed.

"I used to tell the girls, you are so interesting, so intriguing, we should write a book about you."

"It didn't work with me," said Joan.

"No, I had to take you out to dinner a few dozen times first," said Allan.

"Good for you, Joan," said Vivian. "Take these math geniuses for all they're worth."

"Julian won me over with a game of chess," said Nancy, kissing him on the cheek.

"Such a charming man," observed Vivian.

The three couples cuddled, caressed by the sounds of the evening. Allan must have gotten uneasy.

"This river reminds me of the cabin we had when I was a kid. There was nothing to do in the boring little town. One day I went down to the beach and watched the tide go out. It never came back."

"Good one," laughed Oakley, revived from his reverie.

"Yeah, you entered town and we had one traffic sign. Resume speed."

They all laughed. Allan had obviously performed this material in the past.

"My favorite traffic sign is 'soft shoulders,'" said Vivian, massaging Oakley's shoulders.

"Mine, too," said Joan, attacking Allan's.

"And mine," chimed in Nancy, kissing Julian on the cheek.

The levity and intellectual sparing of the evening had functioned as resting his brain before Oakley's final exam. A little more riverside bantering and staring at reflecting and radiating dots of light in the night sky would stretch into the evening, but soon they would hit the tables.

AIN'T NO WOMAN LIKE A ONE-EYED GAHT

15

Toy Slot Machines and Other Diversions

Julian and Nancy excused themselves, but the others returned to Harold's Club to try out the complete point count system. Manny, unlike in the diversionary afternoon session, was intent on watching. He, with Oakley, Vivian, Allan, and Joan, walked north across the bridge.

Now with serious gambling time approaching, Oakley hoped that he would face no difficulty from the pit boss, that he would not be taken as a serious threat to Harold's profits. That was the point of the gambling during the afternoon.

Reno was a good place to try out Julian and Oakley's system. Las Vegas is for vacationers from all over the world, those who are willing to spend a lot of money on lavish rooms, dinners, shows, and gambling, or just to be able to say they were going to "Vegas," implying they were adventurous and oblivious to danger, a modern swashbuckler. South Lake Tahoe draws people from the Bay Area for its skiing, scenic woods and lakes, and romance, with gambling a side attraction. South Lake Tahoe is, strictly speaking, a vacation rather than gambling destination. I think I noted that before.

Reno, though, is a gambler's town. In Las Vegas, unless you're playing downtown, you need a car to go from casino to casino. In Reno, they are side to side on Virginia Street, smaller venues like the Money Tree, Silver Spur, and Primadonna, as well as the big name operators such as Harrah's, Harold's Club, Fitzgerald's, the Cal Neva, the Mapes Hotel and, across the river, the Riverside Hotel. Virginia Street is almost congested, like the stores on a strip mall. It reminded Oakley of the portside merchant stalls in Madras, India, where he had given a paper at the renowned Presidency College at a

© Springer International Publishing AG 2017
L.M. Golden, *Never Split Tens!*, DOI 10.1007/978-3-319-63486-9_15

symposium to honor the great astrophysicist Subrahmanyan Chandrasekhar, a 1930 graduate of that university who would win the Nobel Prize for his work in developing the theory of stellar structure and evolution. Each such stall shared a wall with its neighbors, with a single merchant seated at a chair at its front.

Before returning to Harold's Club, Oakley and Allan decided to warm up at a couple of the smaller casinos. Call them practice sessions. At the small Money Tree, having only two tables, Oakley bet $10 to $100 depending on the value of the high-low index, and within minutes won about $250. He stayed for several more hours and won another $400 or so. Manny, Vivian, and Joan watched. A trip for fun with your girlfriends and, perhaps, your father, makes for excellent camouflage.

Then they walked over to the Silver Spur, for their second and final warm-up. Feeling more at ease with the system, Oakley's betting now ranged from $25 to $250. In fifteen minutes, he rang up another $500 in winnings. Allan was playing the basic strategy at the same table, having not studied the complete point count. That these friends did not make the same strategy decisions provided another effective camouflage.

Nonetheless, Oakley's winnings led the dealer to lean forward over the table and press a concealed button on the floor with her foot, a call to casino management. Two managers appeared, exchanged greetings with Oakley and his friends, and then instructed the dealer to shuffle the deck between 12 and 15 cards from the end. Casinos had learned, over the previous decade, that some players would wait until near the end of the deck and then jump their bets from $1 to as much as $500. Not knowing what the players were doing, but realizing they were cashing in, the dealers would be instructed to shuffle the deck five to ten cards from the end. Management had decided that Oakley was one such player. Oakley would later take great advantage of what he would call "end play."

Yet the ensuing decks were favorable, with positive high-low indices, and Oakley continued to win. Management returned and told the dealer to shuffle 25 cards from the end. Allan now decided to lay out, letting Oakley play by himself. The decks continued to favor Oakley, and he continued to win. Management returned again, and told the dealer to shuffle 42 cards from the end of the deck, after only two hands had been dealt.

Oakley continued to play for another twenty minutes or so. Despite the deck becoming unfavorable, the house possessing some unfavorable table rules, and the early shuffling, Oakley was able to win another $80. With winnings of nearly $2000, it seemed pointless to continue at the truncated-deck addicted Silver Spur.

With hours of practice behind him, Oakley and his entourage of hardened gamblers moved on to Harold's Club for the kill.

Oakley purchased $2000 in chips from the cashier and then he and Allan sat at a blackjack table set against the wall in what could best be described as a corridor between two large gambling rooms. They hoped that they could have the table to themselves, gamblers preferring the glitz of the larger rooms.

During the next couple of hours, much of the time Oakley and Allan did have the fortune of having no other players at the table. Yet Oakley had the misfortune, to use an inappropriate term, of losing steadily. The point count had frequently been negative, leading him to usually bet his minimum $25. Yet, he was down nearly $1700, which nearly wiped out all of his earnings at the Money Tree and Silver Spur. He was, understandably, discouraged. He fought off any thoughts that although the complete point count system was sure to win in the long run, perhaps, in the more realistic short run, streaks of bad luck would make the system non-viable. Perhaps that book that he and Julian had planned on writing would not be written. Maybe he should just publish his dissertation, the traditional manner of advancing one's nascent academic career. It would certainly please Vivian.

Allan was taking frequent breaks, letting Oakley play heads-up against the dealer. When he did play, he was also losing, but, not playing Oakley's stakes, the amount was only a hundred dollars or so. Then things changed.

Oakley had laid down a bet of three hundred dollars in $25 chips. Allan, playing to Oakley's left, had bet $20, four $5 chips. Their two cards were face down. Joan, Vivian, and Manny stood behind them.

Jo, the dealer, had an up-card of the six of Clubs. The only other witness was Elroy, the pit boss, who stood behind Jo watching the action.

Excluding the burned card, which had not been viewed and which was therefore, according to Oakley's complete point count system, not to be included in the tallies, 16 cards had been played out of a newly shuffled deck. The point count before the hand he now held had been +9. The high-low index had therefore been $+9/36 \times 100$, or +25, calling for a bet of 12 units of Oakley's minimum bet of $25, or $300. This was the table maximum at some Reno casinos, a large bet, but not at Harold's, where the table maximum was $500. Betting the table maximum would certainly have attracted significant management attention.

Oakley held a pair of eights, the eight of Clubs and the eight of Diamonds, each of point value zero. Jo's up-card of six had a point value of +1. The updated point count, which Oakley referred to as the "running count," was therefore now +10, and the high-low index was $+10/33 \times 100$, or about +30. In the simple point count system the point count was only updated after all the cards had been played, as a guide to the bet for the next hand. In the complete point count system, the running count, obtained by adjusting the point count as every card was dealt and seen, was kept to guide the play of the hand in progress.

Oakley flipped over his two cards. He separated them, indicating he was going to split the pair. This is a standard bet for anyone with the minimum knowledge of blackjack, sixteen being a lousy hand. Oakley didn't have enough chips on the table to cover the entire bet, but when you split a pair blackjack rules at the vast majority of casinos require you to match the original bet, and Oakley wanted to do so. He was too involved in the hand to even consider playing dumb. He reached into his pocket and pulled out his wallet. He removed three $100 bills and placed them on the table next to the eight of Diamonds.

"Money plays," announced Jo.

She dealt the three of Clubs next to the eight of Diamonds and the ten of Hearts next to the eight of Clubs. With point values of +1 and −1, these cards canceled each other out. The high-low index was now $+10/31 \times 100$, or about +33.

Oakley pulled out another three $100 bills from his wallet and placed them next to the eight of Diamonds. Even the basic strategy directed, as most all players know, that you double down with a hard total of 11 against the dealer's up-card of 6. The complete point count tables that he and Julian had developed said to double down when the high-low index exceeded −35.

Such an exceedingly large negative index would result from, for example, a point count of −7 with only twenty cards remaining in the deck to be played, $-7/20 \times 100 = -35$. That meant, of the twenty remaining cards, at least seven, the exact number depending on the number of ten-value cards or Aces also remaining, would need to have point values of +1, the 2's through 6's. If, for example, four ten-value cards or Aces were present, then the deck would have to have eleven of these low-value cards, the rest being the sevens, eights, and nines, with zero point value. Other than such unfavorable situations, the complete point count system directs the player to double down with a hard 11 against the dealer's up-card of 6.

"Double." Perhaps Oakley was getting carried away. The term for doubling your bet is "double-down." One needs a fair amount of experience to feel comfortable using the abbreviated term, and the whole point of the diversion in the afternoon was to convince the casino that he was not an experienced player.

Vivian was alarmed by the amount of money Oakley had wagered. "$900, Edward," she advised.

"Money plays," Jo announced to the assembled throng.

She dealt the seven of Clubs next to the three of Clubs and pointed to Oakley, a non-verbal query if he wanted to take another card. He had an 18, a standing hand.

Oakley waved his hand over the table.

"Stand." Again, Oakley could have avoided the wave of the hand gesture and simply said "I'll stand" or "I think I'm okay."

Jo pointed to Oakley's hand containing the ten of Hearts next to the eight of Clubs, with the same query.

"Stand." I guess it could be expected, with the money on the table in their first real test of the complete point count system, but a little banter might have led Jo, Elroy, or anyone observing from the one-way mirrors in the sky to think, as desired, that he was not a serious player.

Oakley's hands played, Jo pointed to Allan.

"I'll take a card," Allan said.

Jo dealt Allan the nine of Spades. She flipped over his cards to reveal the nine of Clubs and the seven of Diamonds. When Allan had called for a card, the high-low index had been +10/30 × 100, or still +33, Oakley calculated. Allan should have stood with his 16 against the dealer's 5, the complete point count system directing the player to stand if the high-low index exceeded −34. Oakley was not surprised that Allan had busted.

As I said, Allan had not had the time to study the system, and, in retrospect, Oakley thought, his gambling buddy not utilizing the same strategies was, as it was at the Silver Spur earlier that evening, useful in avoiding detection. He would tell Allan later than he should not have taken the card.

"Bust," Allan moaned.

Jo took the four chips from Allan's position and deposited them in the chip tray. She flipped over her down card to reveal the ten of Hearts.

"Sixteen." For whoever is counting, this was the third lousy total of sixteen that had been played that hand.

Jo dealt herself the eight of Clubs.

"Dealer busts."

She stared at Oakley with a blank expression. Allan stared at Oakley with a smirk on his face. Vivian stared at Oakley, shaking with silent laughter. He had just pocketed $900, cash. That was more than his beginning $7000 annual salary would generate in more than a month and a half.

They weren't aware of the spy cameras in the ceiling peering at them.

The deck was still "hot," the high-low index being +10/28 × 100, or about +35, and, having established this betting pattern, Oakley continued to place bets larger than the minimum. In the one deck he made up his $1700 in losses at Harold's Club.

Oakley knew that winning that amount of money in one deck, let alone one hand, wouldn't be duplicated soon and accordingly decided to take a break. He had experienced what he would experience many times over, moderately heavy losing streaks, followed by dazzlingly brilliant streaks of large winning hands. This had taught him an important lesson. He could not allow himself to be discouraged by the losses to the extent that he lost faith in the system and either lost discipline and stopped obeying its directives or, even more unconscionably, abandoned playing. In addition, suddenly decreasing his bets to the minimum would tip off the casino, at least Elroy, that he might have a viable system. It was time to take a break.

He and Allan stood up from the table and, with Vivian, Joan, and Manny, they left the table and walked out of the corridor into Harold's lobby.

"My linear operator!" exclaimed Vivian, giving her suddenly wealthy husband a hug.

"Not bad for five or ten minutes work," added Allan. "A few more hands like that and they'll have to give you tenure," he joked.

"Yeah, but now you gotta be careful," said Manny, stating the obvious.

"Yeah," agreed Oakley. That kind of action is exactly what draws casino attention.

"Look, I think I better not go on the rounds with you no more," advised Manny. "They know my face. Right now you're just a kid looking for a lounge act. I'll just wait outside."

"I think that's right," agreed Oakley.

"By the way," Joan wondered, "anyone seen Julian and Nancy?"

"Probably taking a dancing lesson," suggested Manny, smiling. "Horizontal mambo or something."

Vivian had seen Manny trying to pull Nancy toward him when they were dancing in the Pair-o'-Dice Lounge. She didn't appreciate the sexual reference and glared at Manny.

"I'll call their room," she said coldly.

Oakley, Vivian, and Joan decided to get a celebratory snack, but Allan, although he wasn't playing the point count system, wanted to continue playing. He had perhaps gotten the bug. He had, in fact, gotten the bug to such an extent that he would himself eventually author a book on gambling. Oakley told him to avoid Harold's, so Allan went down the street to the Cal Neva. After about an hour they met in Oakley's room at the Mapes Hotel.

Allan had been playing blackjack and had an interesting tale to tell his friends. The pit boss and dealer were discussing some guy who had carried what they called a magic slot machine. He always wore a white turtleneck with a gold chain around his neck, which he said was for "good luck."

In any case, Allan's eavesdropping allowed him to fill in the story of the guy with the turtleneck and gold chain. It seems he had gotten a little plastic toy slot machine as a souvenir welcome gift at the Westward Ho motel in Las Vegas years earlier. The toy was actually operational with a functional handle and all the fruit symbols on three separate wheels. Apparently, he would spin the toy slot machine handle and show the result to everyone at the table. If the three columns of fruit came up all cherries he would bet. Otherwise, he would not. The guy, who liked to be called Goose, would win an uncanny number of hands when he bet. When he lay out, the other players would more often than not lose.

Goose, according to Allan's understanding, quickly attained cult status.

At the Four Queens downtown a couple years earlier, Goose almost lost the toy slot. For four straight deals everyone lost, but Goose had lain out. On the next hand, Goose pulled the slot machine handle, showed it was all cherries, laid down a $100 bill and got blackjack. The guy sitting next to him got enraged and grabbed for the toy slot, saying, "Let me see that slot machine."

Generally, if you sit you have to play. If you don't want to bet, protocol states that you should leave the table and give up your seat to someone else. The dealers usually enforce this, though gently. Here, though, was a guy with a good luck gold chain and a seemingly cosmically connected magical toy, and the dealers and pit bosses were as intrigued by it as much as anyone else and they let him play.

Later, walking across Virginia Street, Allan related that he saw a guy wearing a white turtleneck with a gold chain around his neck. On a whim, he asked the guy if his name was Goose. The man was surprised at this question, according to Allan, but admitted he was the same. Goose wanted to know if Allan worked for one of the casinos, and Allan told him that he was a mathematician and that he and another mathematician were testing out a system for blackjack. This apparently opened the figurative floodgates of verbiage, because Goose admitted that he was an electrical engineer, with a bachelors' and masters' degree in the difficult VI-2 honors program from, of all places, MIT, and that he, too, had a system. It called for lying out when the deck favored the house and betting heavily when it favored the player.

Allan asked him about the magic slot machine. He related, with some enthusiasm, as if years had gone by without his being able to share the information, that he had discovered while playing with the slot machine that the slot machine was defective. If he pressed in slightly on the two sides of the toy slot with his thumb and index finger, it would come up a winning three cherries. Without that pressure, the slot would come up other, random, non-winning fruity combinations.

Just like Allan and Oakley were camouflaging their play by pretending to be musicians, as Manny had suggested, Goose was camouflaging his play. He would sit at the table and pull the lever. When the deck was cold, he wouldn't press in the sides, random fruit would appear, he'd display it to the dealer and the other players, and would not bet. Goose had a perfectly rational excuse. "It's bad luck unless the cherries come up," he would say. When the deck was hot, Goose would pull the lever, push in the sides of the magic slot machine, and display for all to see that the machine god had decreed that he lay down a bet. He would nearly always win when the magic slot machine came up cherries, and the other players would usually lose when it did not.

Allan remembered that Clive, the pit boss at Harold's Club, had mentioned that the gambler who called himself the Stargazer used a similar device. On mentioning this to Goose, Goose laughed and admitted that the Stargazer was in fact his twin brother and that a couple years earlier they had traveled through Nevada, gambling together in Reno, South Lake Tahoe, Vegas, and Carson City. Goose took the magic slot machine with him on every gambling trip.

Allan's narrative, in fact, emboldened Oakley. Here was a guy, by all evidence, genuine, who had a system enabling him to distinguish cold from hot decks. It was essentially what he and Julian Braun had developed. He would have loved to have met Goose and to have compared systems, but that was the last time any of them encountered him. Perhaps Goose, on reconsideration, had thought Allan was in fact lying and was indeed a casino employee, and that it was best to leave town before either his photograph was taken for dissemination, which would ruin his game for the future, or he was barred or even arrested for breathing too much air or walking on the wrong side of the street.

*   *   *

Before the proliferation of relatively low cost television monitors and switching circuits driven by computer software, the tables and slot machines were monitored by personnel in the rooms in the ceiling above the casino floor, the security office. They'd be seated in front of closed-circuit television monitors, each person monitoring a few screens. The images were captured by security cameras sitting behind one-way mirrors in the ceilings of casinos. Some referred to them as eyes in the sky. Today, the security rooms can be located anywhere in the building, wiring being able to carry the signals anywhere.

These surveillance systems and the monitoring personnel paid for themselves many times over. They would look for cheaters, both players and dealers in cahoots with players. The personnel, most of whom were former dealers and perhaps former pit bosses, knew the kinds of signals between dealers and players used. They particularly paid attention to high rollers. A gambler betting $500 or more a hand could pick up thousands of dollars on a hot streak, and the casino wanted to make sure that in such circumstances only luck was involved. They'd look for the player marking cards, either by bending them or putting foreign substances on their backs. They'd look for players repeatedly taking hits when they should have stood or standing when they should have taken a card, both indicating that the dealer was peaking at the next card to be dealt and providing the drawing or standing advice to the player. It was a continual game of spy versus spy, the cheating players and the casino trying to stay abreast of each other's techniques.

Sometimes, if a player had won some significant money, the dealer would go upstairs and sit and watch after his shift change. If the player reappeared on another shift, and that dealer was in the casino, he may be called to go upstairs and observe again.

Slot machine players would also be observed. Although historically mechanical devices, more and more were electronic, and these could potentially face tampering by a sophisticated cheat.

Generally, however, life in the sky couldn't be more boring. Most players are unsophisticated. This is particularly true in Reno, which lacks the glitz and high rollers of Las Vegas and the romance and well-heeled skiing clientele of South Lake Tahoe. Reno attracts the unfortunate fringe, who can stay in cheap rooming houses or hotels while trying to make a few bucks in the casinos. Except for the occasional spilled drink or cigarette burn on the green felt, they create little problems in the casinos.

The security personnel would look at the cameras, finding nothing, and munch on doughnuts and pretzels, with a sip of soft drink or coffee. Because of the nature of their work, and its potential importance, booze was not allowed in the security rooms.

If something out of the ordinary were occurring, the security man would call down to the pit using phones sitting on tables next to the closed-circuit television monitors. Today something out of the ordinary was occurring.

A middle-aged man sat before a couple of the monitors, a blank stare on his face. He munched on a cinnamon roll. He picked up the phone and dialed.

The phone on Elroy's table rang. Elroy reluctantly left the dealer with whom he was chatting, and picked up the phone.

"Yeah?" he knew who was calling. The only calls he would get on that phone were from in the ceiling.

"What was that about?"

"I don't know," said Elroy. "This kid presented some tokes from Bernie Kurtzman and he got lucky."

"Bernie Kurtzman?" queried the man upstairs.

"Yeah. I checked. This guy and his girlfriend have a lounge act, and they played some tunes for Bernie to get a gig in the Pair O' Dice. Sounds okay to me."

"Who was the pal?"

"Said he's the manager for the act. I don't know. Sounds okay. Just lucky. They were playing this afternoon and didn't know what the hell they were doing."

The man upstairs was tired of the responses he was getting from Elroy. His job was to keep these types of people off the tables, but he couldn't do it without the assistance of the pit bosses.

"The guy pocketed nearly $2000 betting, what, only twenty-five dollars?"

"Yeah. He was losing steadily for an hour or two then he jumped it to three hundred." Elroy was trying to defuse his supervisor by mentioning the higher level bets.

"He made it all up in one damn deck, Elroy."

"Well, I don't know. He was lucky."

"And what the hell was Manny Kimmel doing there?" He wanted information and was getting nothing he didn't already know, even though he and Elroy had worked together for so long they shared each other's vulgar expressions.

"I don't know," said Elroy. If the man upstairs hadn't worked with Elroy for more than ten years he might have thought he was working with these guys. "Didn't talk to anybody. Just watching maybe. You got this guy's picture?"

"It'll be on the street in ten minutes," the man upstairs replied. He knew that Elroy knew the procedure and he thought the question wasn't even worth asking. "And a couple of the guys are gonna follow him around."

*   *   *

Reunited, the four decided to go back to the Riverside Hotel, where they had had dinner, to continue the experiment. No one there had seen Oakley play, either as a rube or as a serious player. It was now nearly 9:00 p.m. Manny had joined them.

Oakley had been able to find an empty table, which was somewhat surprising. Most of the tables were full, as expected at this time of evening. This most likely meant that the players had been losing there, dissuading others from sitting. Unless the dealer was actually cheating, or the deck was marked, two highly improbable situations, the empty table was another symptom of players' ignorance. They may have thought the dealer, the table, or the deck of cards was unlucky. As Oakley would repeat numerous times, except for the occasional slight rule variations, the game is the same at every table at every casino. It's a good lesson to learn just in case you do run across a marked deck or dealer cheating in any of the ways that Manny had described. Try to get your money back at another table. Don't chase your money, as the gamblers say.

Vivian, Allan, and Joan stood behind the table. Manny stood off to the side. Allan did not recoup his loss at Harold's Club, and Oakley would play alone. He had started out betting $20 minimum bets, and after winning several hundred dollars over an hour, he increased his minimum bet to $25, with a maximum of $250. That wouldn't draw attention, he thought. Oakley then started betting at two positions on the table.

He had bet $75, three $25 chips, at each of the two positions. Playing two hands is not uncommon, but it was another bad decision if Oakley wanted to hide that he was a player with some sophistication. They had played two hands, and thirty-three cards remained in the deck, excluding the unseen burned card. The point count before the deal had been +2, providing a high-low index of $+2/33 \times 100$, or +6. He had accordingly laid down the three unit bets, half the high-low index, $25 being his unit, minimum bet.

Oakley's hands consisted of the ten of Spades and six of Spades, and the nine of Clubs and seven of Diamonds, both totals equaling the dreaded 16. The up-card of the dealer, a young man named Wally, about the age of our group, his straw-colored blonde hair styled into a precise flattop, was the eight of Hearts. The point values of the two Spades canceled each other out, and the two cards in Oakley's second hand and Wally's card all had point values of zero, so that the point count remained at +2. The high-low index was now $+2/28 \times 100$, or about +7.

The complete point count drawing and standing table directed the player to stand with a total of 16 against the dealer's 8 if the high-low index exceeded +11. It didn't, and Oakley had accordingly asked for a card. The dealer dealt the four of Hearts to the ten of Spades and six of Spades.

"Stand," said Oakley. He now held a standing hand of twenty. The point count was +3, and one more card had been played, yielding a high-low index of $+3/27 \times 100$, or about +11.

Actually, the high-low index was slightly more than +11 because 1/9 equals 1.1 to one decimal point. This is how Oakley thought, although he knew it was usually an unwarranted increase in precision relative to the tables he had devised for playing the system. This case was an exception, +11.1 exceeding the high-low index decision value of +11 for determining his strategy for playing his second hand of sixteen.

Wally pointed to the nine of Clubs and seven of Diamonds.

"Stand," said Oakley, using the knowledge of that decimal place.

"You got a sixteen there, too, you know," Wally said.

"I know."

"Strange," murmured Wally.

"I just don't feel as lucky with this one," said Oakley.

Wally flipped over his bottom card to reveal the seven of Clubs.

"Fifteen."

He dealt himself the eight of Diamonds.

"Dealer busts," he announced. "You guessed right."

"Luck," Oakley smiled. Oakley would remember this hand. Not only did he win $150, but the hand had also provided an interesting conundrum. If the pit boss or man in the sky would ever become skilled in the complete point count system, then he would realize Oakley was changing the point count on the fly, adjusting it as every card was dealt and seen, using, as I said, what he referred to as the running count. That would indicate he was a sophisticated player and must be barred.

On the other hand, the inconsistency in dealing with two hands both totaling sixteen, taking a hit on one and standing on the other, might indicate, as it did to the dealer here, that the player was basing his decisions on "feeling," "luck," or some superstition. At this point in his career, so to speak, with virtually no one except him, Julian, Harvey Dubner, Allan, and Manny aware of either the complete point count or hi-lo systems, he had appeared as a player feeling lucky or unlucky. The future might resolve the casino's conundrum in the opposite way.

As far as luck was concerned, Oakley knew that players following his system would be considerably "luckier" than the average player. They might be embarrassingly lucky.

Wally gathered up the cards from the table as Oakley now laid down six $25 chips at each of the two positions, doubling his previous bet. Seven of the last eight cards dealt had been non-tens, and the deck had become more favorable. The point count was +3, and 26 cards were left in the deck after Wally had busted, yielding a value of the high-low index of $+3/26 \times 100$, or about +12. That called for a six-unit bet of $150 at each position

Although Oakley might have been concerned about doubling his bet, in fact he had discovered that casino personnel weren't particularly alarmed at such a strategy. One of the best known "systems" for many gambling games is the Small Martingale, also known as the "doubling-up" system. In it, a player doubles his bet each time he loses. When he eventually wins, he has made up all his losing bets plus his initial bet. He then reduces his bet to his minimum. In this way, say the player loses bets of $1, $2, $4, and $8, but then wins his bet of $16. He has lost a total of $15 on the first four bets, but then cashed in, netting $1. The system earns its popularity from its simplicity and apparent inability to go wrong.

The system fails because houses have maximum bet limits. For a club with a $300 maximum bet, if your initial bet is $20, for example, you need only lose four times in a row, totaling $20 + $40 + $80 + $160 = $300, to reach a level which you cannot both make up your losses and also earn a profit, the $320 bet being disallowed. Even in an even-money game, such as flipping a fair coin, the probability of getting four losses in a row equals 1/2 multiplied by itself four times, 1/16, equivalent to 6%, and no casino game is an even money game. In a real casino game, the probability of stringing four losses in a row is greater.

Worse than that, the Small Martingale provides a bad strategy for determining your blackjack bet size. When the deck favors the player, he or she should be increasing the bet and when the deck favors the house the player should decrease the bet to a minimum, the unit bet. If the player faces a favorable deck, it may result in a succession of winning hands. It makes little sense, other than if some camouflage technique is being used, to decrease the bet to the minimum. Yet, that is what the Small Martingale system dictates. Similarly, if the player has lost a number of hands in a row, it may indicate that the deck favors the house. Again, it makes little sense to be increasing the bet in such circumstances. Yet, that is what the Small Martingale system dictates. The Small Martingale system tells you to increase your bet when the deck may be against you.

This may be an occasional good camouflage, as I said, particularly if a casino employee in the sky suspects you are a card counter and is counting along with you. Generally, it violates a basic premise of systems such as Oakley's complete point count system.

Nonetheless, the system, particularly for players placing small bets at casinos with large maximum bet limits, is popular and, more importantly, easily detectable and well-known to casino personnel as being unsuccessful. Oakley, by doubling his bet after winning, was employing what seemed to be a variation on the Small Martingale system. It was a method of camouflage.

Allan was aware of Oakley's ploy and decided to support his efforts by writing a script and engaging Vivian and Joan as well as Oakley in its creation.

"That's the Small Martingale system that he's using," said Allan. Oakley turned to look at him, but said nothing, indicating that he did not disapprove of Allan's foray.

"This system is ancient, truly ancient," declared Allan, demonstrating for the first time to all of his friends that he was a gambling history buff. "What was to become known as the Small Martingale was created by Pierre Gronk and Francois Klunk, the famous Cro-Magnon casino magnates," he lectured.

To this unanticipated imaginative invention Oakley emitted a deep laugh, knowing that his friend was for some unknown additional reason, perhaps boredom, creating a very tall tale. Oakley continued the narrative.

"Yes, what Allan says is right. Although my colleague Claude Shannon invented the wheel, Messieurs Gronk and Klunk made their fortune in the wheel business and indeed constructed the first roulette wheel. Its image appears as a cave painting in the prehistoric Cave du Dublecero in southern France."

"That's right," said Allan. "Of course, roulette 40,000 years ago was simpler because man then could only count to five, the number of fingers on one hand. The other hand was used in walking."

To this bizarre visualization they all laughed heartily, including Wally the dealer, his flattop hair shaking from front to back like a straw broom. "They developed the Small Martingale system, but they didn't call it the Small Martingale because language hadn't been developed yet, right Wally?" added the well-known linguist Oakley, to additional laughter.

"Right," agreed Allan. "They called it the Gronk-Klunk."

"Makes sense to me," said Joan. "We are truly blessed to be in the presence of such scholars, aren't we, Vivian?" said Joan in mock admiration.

"Sure, fascinating archaeological history," agreed Vivian.

Hopefully Wally felt like one of the gang and would cooperate with Oakley's betting pattern. One could be sure that such entertainment was not common at the tables.

Smiling, Wally dealt two cards to each of Oakley's $150 bets and two cards to himself. His up-card was a six of Hearts. Oakley had been dealt a five of Diamonds and seven of Hearts and an eight of Clubs and ten of Diamonds. With five more cards dealt and seen and the point count now at +4, the high-low index was +4/21 × 100, or about +20, telling Oakley to stand pat with both hands against the dealer's 6. The drawing and standing table directs the player to stand with a 12 against the dealer's 6.

Wally flipped over his down card to reveal a Queen of Spades, and he drew a six of Clubs, busting. Oakley made another $300 bet. The high-low index was +4/19 × 100, or still about +20. The complete point count system called for a ten-unit, or $250 bet. He left the six $25 chips that he had won on both hands next to his two original $150 bets. He was going to, in the spirit of the "reverse" Small Martingale camouflage, "let it ride." It meant a twelve-unit bet instead of the system-directed ten-unit bet, but it was a good camouflage, he thought.

"Let it ride," he accordingly said.

Wally dealt two cards to each position and two cards to himself, but before Oakley could pick up his cards the pit boss walked over. He pushed Oakley's chips back to Oakley and held a photograph next to Oakley's face.

"You take pretty pictures," said the pit boss. He was not smiling.

Oakley looked at the photograph. It had been taken at a very skewed angle, the result of the camera being in the ceiling.

"And that's without make-up." Oakley was shocked but after a swallow had managed to regain his smart-aleck swagger.

"Funny," said the pit boss with false admiration. "We don't want your action here." He really didn't know why this fellow was doing so well, but the report from Harold's Club and his running up almost $500 in less than a quarter of an hour at his casino was enough to get him removed. The "reverse" Small Martingale camouflage hadn't worked, Cro-Magnon inspired entertainment or not. Oakley was beginning to learn not to use a large range of bet sizes, no matter what camouflage he was using.

Casinos are basically private Clubs. As you enter any casino, a large sign prominently displayed somewhere on the wall informs you that the casino can remove players at their discretion, for any reason. Winning a lot of money is a reason.

The pit boss waved to two men who quickly walked up behind Oakley, grabbed him by the arm, and yanked him up from the table, pushing aside his shaking wife and bewildered friends. Manny had skulked away when he saw the goons approach.

"Next time," the pit boss continued, then he thought again. "But there ain't gonna be a next time, right?"

*   *   *

Oakley, Vivian, Allan, Joan, and Manny stood on the sidewalk outside the Riverside. They could see the narrow river, and its placid quality was a contrast to the stress they had just experienced. Oakley buttoned his shirt and straightened his collar. He had been somewhat manhandled and something in his shoulder was ripped, torn, or at least stretched. He rubbed his shoulder.

"Looks like my days of throwing ninety-mile-an-hour fastballs are over," he noted. "Enough income for one day."

"Another reason that martial arts should be a required course for math majors," offered Allan.

"You're a strange person, Allan," observed Joan.

"I don't like this, Edward," said a concerned Vivian. "You're a mathematics professor, not a prize-fighter."

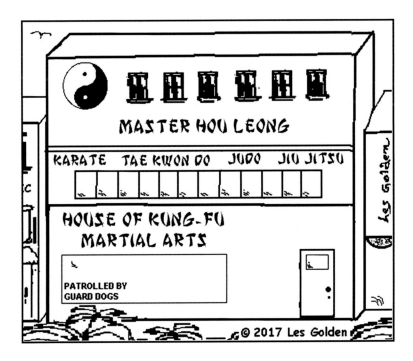

"Until you get tenure, it's about the same thing," philosophized Oakley, trying to minimize the earlier event.

"I actually think it's quite funny," said Allan. "Great story for the faculty lounge."

"Well, I don't think it's funny," Vivian said firmly. "You're risking real physical harm just to verify a computer simulation."

"It's for science," Oakley replied.

"Science is great, Edward, but putting yourself at physical risk is crazy."

"Astronomers risk frostbite and skidding off icy roads to go to mountain observatories," testified Oakley. "Marine biologists risk stingray and shark attacks and oxygen deprivation by going on dives. Geophysicists and climatologists risk starvation and freezing to death by living in Antarctica."

"And Marie Curie died of cancer resulting from studying radium," chimed in Allan.

Vivian ignored him and addressed her husband. "But you don't have to do this. You're living in Massachusetts writing computer programs!" She pulled a piece of paper out of her purse.

"And here's a message left by Julian for us at the hotel. They checked out." She handed it to Allan "Here, Allan, read."

Allan read the note. "Dear Dr. Thorp. Nancy and I are going home. Tell your friend Kimmel, quote, no, Nancy won't accept his invitation to visit him in his room. Yours, Julian."

"You think that's funny, too, Allan?" asked Vivian in a mixture of anger and sarcasm. Edward? I don't think it's funny." She was upset at Manny's treatment of Nancy and his disrespect for Julian, but that was all in the substantial context of her concern for Oakley.

Manny responded. "What's that about? The guy's nuts. I think I was going to show them how to do some dancing, that's all." Then he turned to Oakley.

"Look, we got more important stuff to worry about. Your mug shot, it's all over Virginia Street."

"Whatever," said Oakley. He was discouraged by being barred, Julian's alienation, and now the strife between Vivian and Manny. "Looks like the experiment is over. As much as I tried to disguise my play."

| BAD CHOICE FOR CARD COUNTER DISGUISE | WORST CHOICE FOR CARD COUNTER DISGUISE |

"No no no no no. My ten grand, I want it doubled," protested Manny. "Profit. After expenses."

"How about a real disguise?" Allan suggested.

"You mean Edward Thorp dressed up like Groucho Marx?" asked an intrigued Oakley. "Nose, glasses."

"No, how about Charlie Chan Thorp?" said Allan.

"No, how about this is all over?" said Vivian. She wanted the experiment terminated, and now it looked like they were simply going to try another strategy.

"This might be fun, Vivian," said Joan, not really empathizing with Vivian. "Maybe disguised like Bozo the Clown or, how about Albert Einstein?"

"Honey, I don't think Einstein would be a good choice as a card counter disguise," observed Allan.

Oakley laughed at the concept.

"Whatever. I say 'yeah,' " said Manny. "Yeah. In fact, a good friend of mine, he does the make-up for the showgirls at Harrah's. Tomorrow. We do this."

"You're really serious," said Oakley. He was becoming more than simply intrigued by the concept. They had become sufficiently indoctrinated in techniques of camouflage, the booze hounds, the magic slot machine, the starving musicians coming for an audition, and the playing worthless common systems such as do-not-bust and the Small Martingale to believe that an actual make-up camouflage might be advantageous.

"It's in the name of science, Charlie," said Allan, casting his vote for Charlie Chan Thorp.

16

Hot and Sour Soup

The dressing room for the showgirls at Harrah's shared the opulence of the high-roller suites and the showrooms themselves. A hardwood table covered with a Formica veneer ran along the rose-patterned red wallpapered wall. It had red felt highlights. A mirror hung on the wall behind the table. Above the table were evenly spaced spherical light bulbs encased in metal housing. Cushioned mesh-backed chairs stood in front of the table. A refrigerator was stuffed with snacks and fruit and soft drinks.

Make-up kits, packages of plastic molding compound, and wire rim glasses lay on a make-up table, out of place except for the job at hand. Manny had purchased the wire rim glasses the afternoon after Oakley was banned from the Riverside Hotel. It was about 7:00 p.m.

Showgirls were walking by in various states of costume. Oakley sat in one chair facing the mirror. Allan and Manny stood behind him. Jeffrey Lyle Segal, a well-known Hollywood, Las Vegas, and Reno make-up artist, stood to Oakley's side. He examined Oakley's face.

"You know you have a perfect oval head," said Jeffrey.

"Versus an imperfect oval head?" asked Oakley, bemused.

"I see round heads, flat heads, square heads. But the oval is the perfection of beauty," said Jeffrey.

"You should'a been a showgirl, Thorp," said Manny.

About an hour later, the project was supposedly progressing. Oakley's face was covered with a shapeless plastic gel. Jeffrey wore plastic gloves," full of the gooey material.

© Springer International Publishing AG 2017
L.M. Golden, *Never Split Tens!*, DOI 10.1007/978-3-319-63486-9_16

"You've done this before, right?" said Oakley.

"No," responded Jeffrey.

A showgirl walked by and stared at Oakley.

"Hi, handsome!" she exclaimed. "Wanna party with us?" She was probably simply being supportive. The girls who had showgirl jobs at Harrah's were at the top of their so-called profession. They had formal dance training, were good athletes, and had to be, in addition, as one might expect, tall, trim, leggy, curvy, and possess perfect teeth and complexion and a glowing smile. Any tricks or even minor dalliances with the patrons was prohibited. They had no personality defects that would have led to mean spiritedness. This particular gal probably had been made up many times by Jeffrey Segal and she knew him to be skillful, no matter the present monstrous appearance of Oakley's face.

"See?" said Jeffrey to Oakley. "Don't question the make-up man."

Manny threw in his two cents. "Yeah. Tell the dealer you wanna party with the showgirls. That's another camouflage."

Oakley's shapeless plastic physiognomy had transformed within another hour to one with Oriental features. He wore a thin moustache and wire rim glasses. Jeffrey held a black goatee in his hand as another showgirl walked by on her way to the stage.

"Finally," Jeffrey announced, "two dollars and ninety-eight cents." He applied the goatee to Oakley's chin. Oakley, Allan, and Manny, who had been present during the entire transformation, joined Jeffrey in staring at Oakley's image in the mirror.

"If I didn't know it was you, I wouldn't know it was you," observed Allan.

Oakley responded in a moderately good Chinese accent, "You talk velly funny."

"See," said Manny. "My *lantzman* friend Jeffrey Lyle Segal, he's the best. Nobody'll know it's you. Guaranteed."

*   *   *

Although his transformation was impressive, Oakley decided to avoid the Riverside and Harold's, from which he had been removed, to put it politely. He decided just to walk out of the dressing room at Harrah's through the showroom and lobby and into the casino. To avoid suspicion, his entourage would be left behind. He would be trying the complete point count system alone for the first time.

He sat down at an empty table. The dealer had the deck splayed out over the table and ran his hand above it as Oakley placed two $5 chips that he still had in his pocket from the experience at the Riverside Hotel in his betting square. The pit boss watched.

"Just in on the chicken soup express from San Francisco?" the dealer asked.

"What?" asked Charlie Chan.

"The chicken soup express. Didn't you come in on the bus with the rest from Chinatown?" the dealer replied.

"Oh, right," Oakley, said trying for an Oriental accent. "I had hot and sour soup."

The pit boss walked over and, to Oakley's dismay, pushed Oakley's chips back to Oakley. He held a photograph next to Oakley's face.

"Well, look at this," said the pit boss in mock amazement. "If it isn't our professor dressed up like a Chinaman. Guess you don't know how to take advice."

This guy was not happy that he and his pit were the next intended victims. He must have taken it personally. Oakley realized he at least should not have played the chips from the Riverside Hotel. Although players do play what are referred to as "foreign chips," Oakley should have bought new chips at Harrah's with cash to use at Harrah's tables. The pit boss grabbed Oakley's cheek and ripped off part of the disguise. He shoved it into Oakley's face.

"Here," said the pit boss. "It's lunchtime. Take the rest off. Now." He pointed across the room to the bathrooms. "There's a bathroom."

After Oakley entered the men's room, several brawny men in overalls approached it. One of them placed a placard on the door handle which read "Bathroom Closed for Cleaning" and stood by the door. He had carried a large portable radio with him and now turned it on high. Two other attendants entered the bathroom. They carried neither mops nor bottles of cleaning material.

Two men, no doubt being told that the bathroom was being closed for cleaning, exited the bathroom. Oakley, who had just entered, did not exit with them. Despite the loud music, a careful listener standing near the bathroom would have heard muffled thuds and thumps competing with the music.

After only a minute, a very short time in which to fully clean a large hotel bathroom, even if fully equipped with cleaning material and state-of-the-art bathroom cleaning technology, the two attendants exited the bathroom. The one standing outside removed the placard and they all walked away.

*   *   *

Oakley lay in his bed at the Mapes Hotel with most of the disguise removed. Cold, damp cloths covered his cheeks and chin. Allan and Manny watched as Joan peeled off the remaining layers of Oakley's facial make-up and fake moustache and Vivian dabbed a cloth over his nose.

"This experiment is beginning to make me feel uncomfortable," Oakley said in a mild understatement.

His pain and near-death experience, nearly as frightful as that of Vivian's reaction and the pillow, had made him recall a *Tales from the Crypt* comic book story he had read as an inquisitive teenager in which a dashing, debonair, heart-throb chemistry professor revealed to his new student-bride in their honeymoon cabin in the Alps the result of various laboratory experiments. Such mishaps as an exploding Bunsen burner, the spilling of nitric acid on his arm, and a careless student mixing sodium with water had left him a pulpy disfigured blob, with most of his appendages as well as face being prosthetic. Oakley wasn't quite in that bad of shape.

At least his marriage was, for the moment, intact, unlike that of the comic book chemistry professor, whose horrified bride retreated to the cabin window and fell to her own ghastly disfigurement. Oakley smiled, remembering the end of the love story. The student survived to be reunited with her professor as a crawling mass of broken flesh, knocking on the cabin door, to soon grapple with the professor on the cabin floor, physical equals, in the consummation of their nuptials.

"Those gals. Must have talked about the disguise," Manny guessed, mistaking Oakley's smile for a grimace. "Big mouths. Casino guy heard 'em talking. Who knows?"

Vivian used the cloth to move Oakley's nose left and right. Oakley's smile quickly dissipated as the pain generated a moan. "Sorry," she said. "At least it's not broken."

"Well, I figure the number of hours you've played with the number of hands you've gotten each hour, you've played a couple hundred hands," said Allan. He had stood on the other side of the room, not wanting to see his friend in pain. "That's not enough hands to make the results statistically significant."

Vivian got the message that Allan wanted Oakley to continue playing, nearly broken nose or not. "Allan, will you look at his face?" she exclaimed in disbelief.

"Allan, maybe she's right," Oakley acceded. The stress on Vivian, the alienation of his colleague and friend Julian Braun, the difficulty he had experienced in being able to play long enough to make good money, and now this physical assault, something he had never really suffered growing up in Chicago, may have changed his mind, need to prove himself or not.

"Maybe they're all damn right. It works in principle. Maybe in the real life of the casino . . ."

"Bull, Edward," said Allan, getting as close to swearing as he ever would. Allan Wilson was a skilled mathematician with an interest in games of chance just as Oakley was. He had obtained his Ph. D. at Berkeley before moving to L.A. to take the post-doctoral appointment at UCLA. He knew that this technique had to eventually work. It was just a matter of learning the correct etiquette, managing your money, letting the law of large numbers work its magic, and adopting effective camouflage. "You know that's stupid."

Manny provided some important tips.

"Your problem, Thorp," he said, "is you're betting too high when the deck gets hot. Tipping 'em off. Might as well wear a neon sign on your head. Just lower your big bets."

Allan moved closer to Oakley's bed. Manny Kimmel was providing advice from the years of experience that they lacked.

"And for cryin' out loud, ya' told me you used chips from Riverside at Harrah's. Stupid. Never play foreign chips. They check between casinos, ya' know? And, damn it, when you've won a big bet leave the table. Don't let them examine your play too much. Pretend you're a kid on a paper route. Collect a little here, collect a little there. You'll be okay from here on in."

"Here on in?" Vivian cried. It seemed to her that she was being outvoted by men who couldn't care about her husband's shoulder, nose, or health in general.

"I only got $2500 of profit. I didn't fly out here for $2500 of profit," said Manny. Then, to use the trick he had employed when visiting Oakley and Julian in Brookline, he went into gangster threat mode. "My investor friends. They won't be too happy."

Oakley was in an unenviable position. Play, and the house goons beat you up. Don't play, and Manny Kimmel's guys might come up from New Jersey and, well, he didn't want to think about it. He certainly didn't want to become the protagonist in a story in *Tales From the Crypt*.

Allan knew the situation. "When are you due back at school?" he asked Oakley.

"Careful!" screamed Oakley at Vivian. Maybe she had tweaked his nose on purpose to remind him of the pain, or maybe she just stiffened up and transmitted that squeeze on hearing Allan's question and its implication.

"I've still got ten days," answered Oakley.

"We gotta get back tomorrow," reported Allan.

"Then you and me, Thorp, we go to Tahoe," said Manny. "Too much cheating in Vegas. Forget about Carson City. Even worse cheating. Lawless."

"What!" exclaimed Vivian. "Tahoe? More gambling? Edward, please."

"Vivian, theory's not enough," said Allan.

"Why don't you do it then?" she demanded of Allan.

"Allan, I don't know," said Oakley. Maybe he had already made up his mind to go to Tahoe and was just mollifying Vivian. Maybe he had decided to give up, although anyone who knew Oakley, including Vivian, doubted it. He was too competitive. "The playing is over here. They know my face."

"Oh, really?" observed Vivian. "They know your face? At least it's almost still in one piece."

"Ed?" Allan notified Oakley that it was time to make a decision.

Vivian, Allan, Joan, and Manny stared at Oakley. He took a deep breath.

"Vivian, I'm going to write a book about this. It's significant science. And I've got to show it works."

"What!? What!? What about your health, Edward?"

"Hazards of the profession," noted Allan.

"Shut up!" Vivian yelled at him. She was not happy.

"What about your dissertation, Edward? I thought you were going to publish your dissertation."

"There's plenty of time to show I can walk on clouds," he noted, showing his obsession both with his father's commentary and his need to prove his detractors wrong.

"You're going to Lake Tahoe, then?" said Vivian, resigned but with an implied threat.

"I have to, Vivian. We need more statistics. You know I have to."

"And me and my nieces, we want a longer vacation," said Manny. He had only admitted to Nancy what everyone knew. "A working vacation."

17

Get a Job

Harvey's Wagon Wheel Casino
South Lake Tahoe
Stateline, Nevada

Oakley would have preferred playing heads-up against the dealer, but at 9:00 p.m. he was happy to have found a table where only two others were playing. It had taken him and Manny, and the girls, about an hour to drive the sixty miles south on Highway 395 from Reno, and Oakley was eager to gamble.

During the ride Oakley and Manny discussed what he had seen and learned. Oakley had observed the dealers staring at his eyes as he counted the cards. He learned quickly to use his excellent peripheral vision, looking anywhere but toward the cards as either players busted or the dealer scooped them up after all hands had been played. He also had to find a way to count rapidly. He found that it was quicker and easier just to cancel out images pair wise, a deuce canceling a face card, for example, than laboriously adding plus and minus ones. In this way the point count could be adjusted by simply adding or subtracting the remainder, usually of an absolute value no greater than two or three.

Oakley, who had learned the power of the right mode of brain processing as a Cornell student preparing for exams, thus realized that he could use the right mode of brain processing to count the cards. He could count more quickly by holistically viewing the images of the cards rather than using the analytical, left mode of brain processing, that is, the sequential counting of them by adding and subtracting plus and minus ones.

© Springer International Publishing AG 2017
L.M. Golden, *Never Split Tens!*, DOI 10.1007/978-3-319-63486-9_17

As a sort of corollary to this, Oakley learned not always to sit in the seventh and last seat at the table, third base. Sitting there certainly provided him an advantage. He could see the cards in the hands of players who had busted, enabling him to keep an updated point count, the running count. If, however, he played for an extended time at one table, or if he revisited that pit on another day, and continued to win, the sitting in third base might indicate to the house that he was using a viable system. This was particularly true if they observed him staring at the cards.

On the other hand, many unsophisticated players believe that the player sitting in third base can control the outcome of the hand, which can result in their becoming the target of hostility. Let's say the dealer has a six showing. If the third-base player takes a hit and gets, say, a face card, the other players who had not busted might get upset, irrationally, that that face card was "meant" for the dealer and could have busted the dealer. Many players, to avoid these types of situations, avoid sitting in third base. That Oakley would in fact sit there routinely might, again, indicate to the house that he was not a naïve player. It was best, Oakley, learned not to sit in the last seat every session, despite it providing him valuable information.

On the analytical side, he was now creating a log of his play. On a piece of paper he would write the name of the casino, the time he played, how long he had played, and his winnings or losses. He included a comment column to note extraordinary events, if a mechanic had been put at his table, if the dealer was particularly accommodating by dealing to the very end of the deck, for example, if he had won a significantly large hand, if he had made the attention-attracting move of splitting ten-value cards, and so on.

The log could serve various purposes. Most importantly, he could track his net winnings. They had refined the paper route technique to directing the player to leave a casino after no more than 20 or 25 units of the minimum bet having been won or lost, about four of five times the conservative maximum bet. More than 20 might attract attention. Less than 20 might indicate he had encountered a cheating dealer. Making the limit quantitative would avoid emotional decisions, either greed or wanting to "get my money back." In any case, he would now leave the table after an hour to avoid even a minimum of scrutiny. Of course, Oakley would be free to modify the criteria for staying or leaving a table based on particular circumstances, leaving if, for examples, a mechanic dealer suddenly appeared at his table or he had split tens multiple times on a single hand and won.

The log also enabled Oakley to avoid other modes of generating excessive scrutiny and detection, leading perhaps to getting barred or again slammed up in some men's room. He would avoid playing the same casino during a given shift. He would avoid playing a casino where he had done especially well on consecutive days or consecutive shifts. In short, he was realizing that to successfully use his system he had to be not only a good counter and employ camouflage but to be a good, so to speak, businessman. He called it "money management."

New and refined camouflage techniques had become apparent. Although he had learned early in his playing to avoid large swings in his bets, he had kept the range of bets from one to at most ten units of his minimum bet, a range of $25 to $250 at the Silver Spur, for example. It would be ideal to bet the table maximum when the deck became favorable and reduce it to $1 when the deck became unfavorable, but such extremes would quickly alert the casino if you were winning. After Manny's comments and his experience at the Riverside, he was going to consider limiting his maximum bet to only three, four, or five units of his minimum bet, particularly when a larger bet would put him close to the table maximum.

As additional camouflage, sometimes he would begin a fresh deck with a large bet, reduce his bet when the deck became favorable, and increase it when it became unfavorable. He would do this more frequently when a pit boss was watching. More than once, after such a contrary move, the pit boss would simply walk away, to Oakley's great satisfaction. Again, if he were concerned that his betting and playing strategies might draw attention he would play at tables with one or more high rollers. The attention would be paid to them.

He had faced at least one cheating dealer, which reinforced his concept that the game is the same, except for some occasional rule variations, at every table at every casino, and not to remain at a table with a cheating dealer to, as I said, "get my money back." Oakley had soon learned that the shifts at a given casino are the same length, typically thirty or forty minutes with a twenty minute break. All the dealers in a given pit would be relieved at the same time. If a dealer suddenly appeared at his table, while the other dealers remained at their tables, it was likely a cheating dealer sent to retrieve the money he had taken from the casino, possibly one of the mechanics that Manny had described at the joyous meeting at the Oakley's in Brookline.

As a partner in the development of the complete point count system with Dubner and Julian, Oakley had faith in it. Although the number of hands he had played, as Allan had pointed out, was yet insufficient to create statistically significant results, Oakley felt that the complete point count worked and that with a longer run it would be validated and he would win, even enough to make Manny happy. Nonetheless, the frequent losing streaks, which might discourage a normal player and lead him to quit, reminded Oakley to remain dispassionate and focused.

He had been playing at a table at Harvey's Wagon Wheel with two other players for nearly an hour, betting $25 to $100. He was up about $400, only sixteen units of his unit minimum bet. The high-low index had been negative a disproportionate amount of the time, and the other players, of course not varying their strategy, had been losing steadily. Neither of the paper route technique signals for leaving the table had occurred, and he was confident that little or no attention was being paid to the table, none of the players being particularly successful.

After the dealer shuffled, the other players bet lightly, trying to hold on. The player to the immediate left of the dealer, Darlene, according to her name tag, had bet a single $5 chip, and the player to Oakley's right had bet two $5 chips. He was sitting in third base.

Oakley had bet $150, laying down six $25 chips. Although this was the first hand being played out of a freshly shuffled 52-card deck, Oakley had placed a bet of six of his minimum $25 bet as a camouflage technique. He was using the camouflage of avoiding starting each new deck with a minimum bet and then increasing it systematically if the deck became favorable. To his chagrin, he had been dealt the ten of Diamonds and four of Clubs. This was a horrible hand, and he would soon see that the other two players shared the same fate. Nonetheless, Darlene's up-card, the six of Clubs, gave them all hope. Manny stood behind Oakley, watching.

With three cards played and viewed by Oakley, excluding the unseen burned hole card, 49 cards remained. The point count was +1. The high-low index was therefore $+1/49 \times 100$, or about +2.

Darlene was a cute, sweet, freckled girl, about 22 years old, young for a blackjack dealer. She shared her name with one of approximately, Oakley estimated, 100 million American girls now being named after the cute, sweet, freckled Mouseketeer. She seemed to be enjoying the background piped-in music at Harvey's Wagon Wheel, "Get A Job," a hit tune of 1957 by The Silhouettes, as she dealt. She would bob her head up and down to the tune. If not for her age and obvious lack of experience, the streak of unfavorable decks might have led Oakley to think she was cheating.

She dealt a ten of Hearts to the first player, who immediately turned over the cards he had been dealt, showing them to be the seven of Clubs and eight of Hearts. He busted, but, more importantly to Oakley, he had taken a hit with his total of 15 against the dealer's 6. Oakley thought to himself that such a strategy move violated even Baldwin's basic strategy. Not only did the player, by taking a hit with 15 against the dealer's 6, violate both Baldwin's and his revised basic strategy, many of whose strategy decisions should be intuitively evident to an experienced player, his play was also an egregious violation of

Oakley's complete point count system, of which, of course, the player had no knowledge. The high-low index had been +2, both of the player's cards having a point value of zero, and Oakley's tables for drawing and standing for hard hands said stand if you have a 15 against the dealer's 6 if the high-low index exceeds −28.

"Sorry," dimpled Darlene told the busted player. She removed his $5 chip from the table and deposited it into her money tray and then scooped up the seven of Clubs, eight of Hearts, and ten of Hearts and placed them in her discard pile.

The player watched and sighed. "You're too hot for me, honey." Oakley nearly gagged as the player left the table. He had a very fleeting urge to tell the guy to read Baldwin's paper. Six cards had been played and viewed, excluding the unseen burned hole card, and 46 cards remained. With Darlene's dealing the ten of Hearts, the point count was now zero, reducing the high-low index to +0/46 × 100, or zero.

The second player scratched the table with his cards, directing Darlene to deal him a card. She dealt him the nine of Diamonds. He, too, busted, flipping over his down cards to reveal the nine of Spades and six of Clubs, another fifteen. He, too, Oakley thought, had violated the basic strategy as well as the complete point count.

"Bust again," said Darlene. "It's the cards." She probably did feel some compassion for these fellows.

"Yeah. The cards," the player agreed.

Darlene took his two $5 chips and deposited them into her money tray, then scooped up the nine of Spades, six of Clubs, and the nine of Diamonds, and placed them in her discard pile. With three more cards dealt and seen, including the six of Clubs, the point count was, assigning zero to the sevens, eights, and nines, +1 and 43 cards remained to be played. The high-low index was therefore +1/43 × 100, or about +2. Oakley's tables directed him to stand with a 14 against the dealer's 6 if the index exceeded −17. If the other two players had been dealt or had drawn a number of ten-value cards or Aces, the index might have gotten that low, but they hadn't been so blessed. Oakley, by sitting in the last chair, had been able to keep a running count, and governed his strategy by this up-to-date count and index.

He waved his hand over his cards, signaling he was standing.

"Fourteen," he said, flipping the cards over. "Gotta stand."

Darlene hadn't been dealing for more than a year. If the two other players had taken hits with their fifteens, it seemed to her that another player should take a hit with the poorer hand that Oakley held.

"You want to stand with a fourteen?" she asked.

"Yep," he said. "You're just too hot," echoing the lingo and sentiments of the other two players, another element of camouflage.

"Okay," she said. She flipped over her down card to reveal a Jack of Spades. The count was now zero, with 42 cards remaining to be played.

"Sixteen." She dealt herself the King of Diamonds.

"Dealer busts," she announced. That, of course, would have been Oakley's card, but he had elected to stand. This was not lost on either the second player or Darlene.

She took six $25 chips from her money tray and placed them next to the six $25 chips in front of Oakley. The deck had favored the players during the very first hand. It would continue to do so during the entire deck. Having established his opening bet at $150, Oakley was able to increase his bet to the $300 without attracting attention. In the rest of the deck he made another $1000. Once again, persevering through long periods of moderate losing was rewarded with a burst of stunning wins.

He was now up over $1700 at Darlene's table, betting generally no more than $100 per hand.

Harrah's Casino
South Lake Tahoe
Stateline, Nevada

After another half-hour of play, Oakley's winnings oscillated up and down, and he left Harvey's Wagon Wheel up about $2000. The paper route technique dictated that he leave the casino before the pit boss or, more likely, any camera-assisted eyes might detect a pattern in his play. It did not dictate that he leave the table in the middle of a favorable deck. That he had begun that highly successful deck with an increased bet was sufficient camouflage to allow him to stay the additional half an hour.

He and Manny moved the operation to Harrah's. Harrah's, of course, was where he was almost murdered by Vivian. He remembered it fondly. As he entered the casino, he thought that he now had the opportunity not only to get that room rental back, but also to get revenge on their terminating his bachelorhood.

He found a table where only one person was sitting. After an hour of play, he was down about $50. He had been betting at the same level as at Harvey's, between $25 and $250 per hand. Manny had wanted to double his $10,000 stake, and Oakley knew that South Lake Tahoe might be the last gambling location where he had a chance of delivering. He decided to increase his betting range from $50 to $300 and also to sometimes bet at more than one betting position. Lady Luck, so to speak, began to favor Oakley. After another hour, he was up over $1500. That was equivalent to 30 units of his minimum bet, and he had been at the same table for two hours, both paper route signals for leaving the table. He decided to remain. He couldn't return until the next shift or even the next day to Harvey's and the number of venues in South Lake Tahoe was limited.

They started a new deck, and Oakley was about even until about 25 cards, half the deck, had been played. The point count was +3 and the high-low index was +3/25 × 100, or about +12. Oakley had bet $300, six of his unit bet of $50, twelve $25 chips.

Oakley had been dealt the Ace of Clubs and six of Diamonds, for a soft 17. The dealer's up card was the five of Clubs. Only twenty-two cards in the deck or, because of the other player's cards, remained unseen, and the point count was +4. That provided a very favorable high-low index of +4/22 × 100, or about +18.

Some kind of Herb Alpert music was playing in the background. Oakley didn't particularly like the lilting, light, pop style of his Tijuana Brass. He thought of the audition at the Pair-o'-Dice Lounge and Bernie Kurtzman's comment about their crowd being a "Yellow Rose of Texas" kind. Oakley would have preferred Maynard Ferguson, a jazz trumpet player's trumpet player. The music of Miles Davis would have been too morose for a casino.

The dealer was a woman by the name of, and I am not making this up, Doreen. Perhaps Darlene and Doreen's mothers had met years earlier in a time-warp at a Mickey Mouse Club fan club meeting. Perhaps, Oakley thought, Doreen's mother had skills of precognition. He would have liked to have met her and tested her for such skills. She would have made a valuable blackjack consultant.

The other player scratched the table with his cards, and Doreen dealt him a King of Hearts. He flipped over his two cards, showing the ten of Hearts and six of Spades. He had bet two $10 chips.

"Bust. Sorry," said Doreen. She removed his cards and chips from the table as Oakley, now in missionary mode, thought he, too, could use a copy of Baldwin's paper. A 16 against the dealer's up-card of 5 calls for standing with a high-low index greater than −34. The high-low index had been +18, in another arithmetical universe from −34. If he had been using the complete point count, an impossibility of course with Oakley only trying it out for himself, his running count and the calculated high-low index without seeing Oakley's cards would have been the same that Oakley had found, +4 and about +18, respectively.

With the dealing of the King of Hearts, Oakley's running count resulting from seeing the other player's originally dealt cards was now +3 and the high-low index was +3/19 × 100, or about +16.

Oakley took six $25 chips from his pile and placed them next to the six $25 chips at his position. Manny, standing behind Oakley, gasped quietly. That was a $600 bet on the table. Only at Harold's Club had he had more on the table, but that had included a split.

"Double down," Oakley announced.

"You don't double down and ask for a card when you got her beat, pal," the other player scolded Oakley.

"Really? Ah. A rookie mistake," Professor Thorp replied. In fact, with a +16 high-low index, Oakley's complete point count system directed the player to double down with a soft 17 against the dealer's up-card of 5 if the high-low index exceeded −28. Professor Thorp knew that +16 was in fact greater than −28, something I would guess he remembered from third grade, or maybe had discovered for himself while resting from calculating the number of seconds in a year.

Doreen dealt an eight of Clubs next to the Ace of Clubs and six of Diamonds.

The other player took a few minutes, it seemed, to tally Oakley's hand. "Now you got a rotten fifteen," he deduced. "Some guys don't know diddly about this game," he told Doreen. Maybe he wanted to date Doreen and was trying to impress the Mousketeer with his acumen.

"I stand," said Oakley looking sheepishly at the player.

"You only get one card when you double down," Doreen instructed him.

"Oh, sorry," said Oakley as the other player buried his head in his hands in disgust. Oakley had learned to make beginner's mistakes every once in a while as a form of camouflage.

Doreen flipped her down card to reveal a Jack of Clubs.

"Sixteen," she said. She dealt herself a King of Diamonds.

"Dealer busts."

Oakley had won $600 on the soft doubling down. He had taken Harrah's South Lake Tahoe for nearly $2000. This was a fine gambling mecca, Oakley thought as he got up from the table. Twenty units of a minimum $50 bet was equivalent to $1000. His $2000 indicated that it was more than time to leave this casino according to the paper route technique that he and Manny had refined.

Barney's Casino
South Lake Tahoe, Nevada

Manny and Oakley left Harrah's after, as Manny had directed, cashing out and leaving his chips there. To Manny and his great pleasure, he had won

$1955. The $2500 he had won in Reno had now increased to almost $6500. A mathematics education is a wonderful thing, Oakley thought as they continued their paper route at Barney's. Madeleine and Renee wanted to tag along, and Oakley thought they would provide good camouflage, two men with their friends partying in Tahoe.

Walking over, Oakley recalled the excursion several years previously when he, Vivian, Allan, and Joan had tried out Baldwin's group's system and had scored some free meal tickets, while encountering the renowned Mr. Marcum. He was sure that, if anyone remembered him, they would do so with a chuckle rather than fear of having Oakley take their money. He went to the cage and bought $2000 in chips.

To Oakley's chagrin, country music was the choice of the vice-president in charge of piped-in ambiance. It was quickly dispelled as Oakley spotted a 52-card deck of cards lying spread out neatly in a semi-circle, face-up, on a blackjack table near the entrance to the casino. No player was at the table.

The dealer, a man about 40 years old, stood with his arms crossed behind the table. A pit boss watched as Oakley and Manny walked up to the table.

"Is this table open?" asked Oakley.

"Sure," said the dealer. He scooped up the cards as Oakley sat down and placed one $25 chip on each of two betting squares on the table. Manny watched as the dealer dealt two cards face down next to each of Oakley's betting squares and two cards to himself, one face down and one face up.

Before Oakley could move his hand to turn over his first hand, the pit boss walked quickly over and placed his hands over the cards. He stared at Oakley.

"If it isn't the Reno movie star," the pit boss said, throwing a photograph of Oakley on the table over Oakley's chips.

Oakley looked at the photograph. "The resemblance is uncanny," he said, being, yes, a smart-aleck. Oakley hid his discouragement well.

"You come back here in your Chinaman outfit, you might find yourself in a terrible skiing accident. You understand?"

Oakley, Manny, Madeleine, and Renee left the casino and sat at a table in Barney's coffee shop. Normally, if they had planned on continuing to gamble, they would have gone to a different casino for a snack or a break. It didn't matter who saw them or overheard them now. Oakley's gambling in the entire state was, at least for the time being, apparently over.

"This ain't good," commented Manny in what was more than a mild understanding.

"What's not good, sugar?" asked Madeleine, who apparently hadn't had her fruit for the day.

"Shut up!" exclaimed Manny, upset at Madeleine's ignorance. "Some skirts, they just don't know when to shut up. Now, Renee, here," he said to Oakley, "she got better upbringing."

Renee placed her head on Manny's shoulder.

"Yeah. I got class. Right, Manny?"

Oakley had been listening and thinking, but was as discouraged as Manny. "What do we do, Manny?"

"It's over here," he said. Then a flash of revelation illuminated his face. "But, you know, they got tables in Puerto Rico, too, you know."

"Puerto Rico?" remarked Oakley, taken completely unawares by this suggestion. He had never entertained the possibility, and the implication of leaving the U. S. mainland was overwhelming. "I want to go see my wife," he said.

Manny dismissed the statement with a wave of his hand. "I say Puerto Rico."

"Yeah, honey. Puerto Rico," said Renee, in a familiarity strange between an uncle and niece.

"I'll be good, sugar," promised Madeleine, using another perhaps inappropriate term of endearment.

Manny essentially ignored them, knowing that persuading Oakley had the higher priority. "The rules are different. Two decks, and they deal 'em face up."

"That's interesting," said Oakley. He envisioned being able to take multiple hands and keep a point count that would account for all the cards that had been dealt, rather than just those he had been able to see on the table as they were flipped over after a hand was played.

"They deal from a shoe, so that makes it tougher to cheat. And you can only double on eleven," Manny continued. "They got a $50 bet limit, but you can bet all the positions on the table, so it don't matter. And the dealers, they deal much quicker than in Nevada, so you get a lot of action."

"That's very interesting," said Oakley. He had considered betting two or maybe three hands, but now Manny Kimmel had told him he could in fact bet all seven positions. If, for example, three-quarters of the 104-card double deck had been played, leaving only 26 cards, he could take seven hands and, counting the dealer's two cards, only ten cards would remain. The point count would give him an excellent idea if they were high-value or low-value cards. He could double down and split to even get down to the very end of the deck. Oakley by this time had had enough experience with the complete point count system to know that what he would call "end play" could be exceedingly lucrative. It was too enticing to pass up.

"I'd have to get Julian to program in the differences," he said to Manny, sold on the Puerto Rico trip. "See the effects of the double deck and the doubling down only on eleven. It might take a couple weeks. But that would be very interesting."

Oakley realized that playing the complete point count system using a double deck and playing with the severely restricted rules for doubling down would provide a powerful test.

"And your picture, it ain't gonna be all over the streets there," said Manny, sealing a deal that had already been sealed. "You don't play stupid down there, you'll be okay."

"Yeah," said Oakley, remembering the beatings, mild and severe, that he had taken, and beginning to think of palm trees and exotic beaches.

"And you grease the palms a little, they let you play," said Manny, aware that Oakley had been effectively banned from a number of casinos from Reno to Tahoe, but not aware that Oakley's mind was wandering to foreign shores.

"Yeah," said Oakley perfunctorily, still considering the scientific, and personal, benefit of successfully trying the system under a different set of rules. After all, success would result in an undeniably total dissipation of the cloud created by his father's disparagement.

Oakley left Manny and his nieces and walked out of the restaurant and over to the bank of phones hanging from the wall in Barney's lobby. He had the hotel operator call Allan's house and asked her to charge the call to his room. As soon as he picked up the phone, Oakley asked him about Vivian.

"She's staying with Joan," replied Allan. "They're going to open up that boutique they've been talking about."

"Does she talk about me?" Oakley asked.

"Does she talk about you? Yeah. All the time. When she's not crying."

"Well," said Oakley. "What does she say?"

"What does she say? She wants to know if your experiment with Manny Kimmel and his nieces is over. . . . Well?"

"It's not over," Oakley answered. If Oakley indeed had the transcendent, supernatural powers that Vivian once ascribed to him, or if he had been less preoccupied with Vivian and more aware of Allan's speech pattern, repeating Oakley's question before answering it, he might have surmised that both Vivian and Joan were sitting on the sofa in Allan's living room. Allan was standing near them, holding the phone, providing Oakley's question for their benefit. Joan was holding Vivian's hands.

Allan shook his head and mouthed to Vivian and Joan, "No." A discouraged Vivian buried her head in her hands. She had hoped that Oakley was calling to tell Allan that he was returning home to Brookline.

"Are you going back to Reno?" Allan asked.

"No," replied Oakley. "My face is too well known."

"So where are you going?"

"Somewhere else," Oakley hadn't sensed the presence of Vivian but the worst thing he could do now was to tell Allan he was actually going, if not to another country, then off the mainland to the commonwealth of Puerto Rico to continue the experiment.

"I see. Somewhere else," Allan informed the girls while speaking to Oakley.

"Allan, give me Joan's phone number," said Oakley.

"You want their phone number?" repeated Allan while looking at Vivian for her approval or disapproval.

Vivian vigorously shook her head "No." Oakley would have to know she was upset, even if indirectly.

"Ed, Vivian doesn't want contact with you right now," replied Allan.

"She's upset. Let her cool off. She needs time."

18

Strike When the Deck Is Hot

The MIT math club met in the mathematics department library, a more congenial location than a lecture hall. The goal was to attract students to an enticing environment, not to subject them to what they might consider to be another course.

The vast majority of book and periodical holdings were housed in the Hayden Library on campus, but the math department library had a couple shelves of the most frequently referenced books, current periodicals, books that the Hayden Library thought were obsolete, and books provided to the department by retired or the families of deceased professors.

A tray of cookies, tea bags, a pitcher of hot water, cups, and a coffee pot sat on a circular wooden table to the side of a blackboard. Instead of classroom desks, comfortable plastic chairs with curved rather than vertical backs were located randomly around the room. Nearly every seat was taken by the twenty students and a few faculty attending. The blackboard was full of equations, tables of numbers, graphs, and a notice at the top of the blackboard which read

MIT Math Club
Prof. Thorp. Probability Approach to the Card Game 21
Next Friday: Prof. Dassoff. Whither Conic Sections?

Professor Ted Martin, the chairman of the math department, sat next to a couple of students. Oakley leaned against a lectern just in front of the blackboard.

"Professor Thorp, you say we should increase our bets if the deck gets hot. But it seems to me that will alert the dealer, and they'll just shuffle the deck on you," said one of the students.

"That's a very good observation, Rick," said Oakley. "That's why I suggest you limit the size of your bets to perhaps no more than five times your minimum bet. That was actually the idea of a friend of mine. And sometimes to hide that you're playing my system, bet a large bet at the beginning of a fresh deck. You could even bet a large bet every once in a while when the deck gets cold, just to throw them off the track."

"Yes, Flash?" he said to another.

Oakley knew the names of all his students. With only about twenty students in the math club, Oakley had no trouble learning their names. He also made a point of learning the names at the beginning of semesters of all the students in his classes. It was easy to do in the relatively low enrollment upperclass courses, but it was more difficult in the large enrollment introductory calculus class.

One student had asked him if he was indeed trying to learn the names of all the students, and Oakley told him he was trying. He explained that this is MIT, and that some are sure to become Nobel Prize winners, members of the National Academy of Sciences, Fellows of the American Physical Society, Fellows of the American Mathematical Society, and, even better, his future colleagues wherever their respective space-time world lines might take them, perhaps even back to MIT.

"Professor Thorp. How about just waiting until the deck gets hot to enter the game?" he asked.

"Great. Such bright students here. Yeah. I call that 'strike when the deck is hot.' It limits how long you can play, of course. But here's the idea. Stand behind the players' backs and watch the play. Count the cards and wait until the count gets favorable. Then quickly sit down and bet a large bet. You're starting with a hot deck, and the dealer has no idea that this is for you a large bet because you haven't established a betting pattern."

One of the students laughed. "That's great," he said.

"Isn't it though, Mr. Ferguson? It's essentially countering the casino shuffling when the deck is hot by not playing when the deck is cold."

"But it's a great way to make some quick lunch money or pay for your lift tickets," said Flash.

"Right. En route to the restaurant, stand back behind a table until the deck gets hot and then make your strategic strike so to speak."

"It's like jungle warfare," noted Ferguson.

"That has occurred to me, Mr. Ferguson," answered Oakley. "We use camouflage, sudden surprise attacks, and we try to blend in with the non-system players."

"Someone should write a book about that waiting until the deck gets hot strategy," Ferguson said.

"Why not you, Mr. Ferguson? There's a similar team strategy I've thought about. The members sit at tables laying down small bets. Then when the deck gets favorable, they signal another member to sit down and begin laying down large bets."

"There's a spy at the table," offered another student.

Oakley laughed. "Right, you could be that spy, Ivan Ivanovich Bobchinsky. Or, the members of a team are sitting at a table and when the deck gets hot they as a team raise their total bet, but to avoid detection some of the individuals keep their bet the same or even lower their bet. Okay, unless there are any more questions."

The students and faculty applauded, rose from their seats, and exited the library. Professor Martin approached Oakley.

"Ed," he said, "that's an interesting application. Come and see me in my office tomorrow, okay?"

* * *

Normally, the question and answer session at the end of a math club meeting lasts shorter than it did after Oakley's "Probability Approach to the Card Game 21" presentation. It was now almost 5:30 p.m. and Oakley was anxious to call Allan again. It was Friday, about 2:30 p.m. in L.A., and Oakley knew that Allan often left his office at UCLA early on Fridays.

Oakley placed his briefcase on the desk in his office and sat in the swivel chair next to it. He picked up the phone and dialed Allan's home phone number. He was right. Allan was home.

Allan sat on his living room sofa holding the telephone.

"I don't know how she is, Ed. She spends all of her time at the boutique with Joan."

"All of her time?" Oakley probed.

"Yeah."

"Nights, too?" Oakley probed further. He was basically secure in his relationship with Vivian, but their separation, even though it had only been a few days, and the recent turmoil in their lives, would have created doubt for the state of any relationship.

"Yeah," Allan said. He knew the anguish his friend must have been feeling and needed to reassure him. "No need to worry about what Vivian's doing at night, Ed. I couldn't tell you the last time you called because they were here, but Vivian picked up a singing gig at Dick's At The Beach."

"They were at your place?"

"Yeah," Allan admitted. "I'm in the middle of this."

"I understand," said Oakley, knowing the situation Allan must be in, cornered by Joan and Vivian. "And she's singing at Dick's At The Beach?"

"Yeah. So during the day, they're at Joan and Vivian's boutique. They call it J and V's Design Shoppe. That's S-H-O-P-P-E, of course. At night, they're hanging out at Dick's At the Beach. I never see my girlfriend anymore because of you."

"What nights are they there?" asked Oakley. He knew that Allan's comment was half in jest.

"Every night," Allan informed him. He wanted to provide further assurance to his friend. "Every single night, Ed. Except July Fourth or if there's an earthquake."

*   *   *

It being Friday when Ted Martin asked Oakley to see him in his office, by "tomorrow" he meant Monday. Perhaps he had in fact forgotten it was Friday. In any case, Oakley spent a difficult weekend alone. He was content that Vivian was occupied, doing what she loved, and not meandering so to speak, but the thought of the looming conversation with the chairman of his department was one from which Oakley couldn't escape.

With their classes and office hours absorbing the day, Oakley didn't make it to Professor Martin's office until late in the afternoon. Martin had to leave so they decided to have their meeting while walking across campus.

"The Moore Instructorship, as you know, is normally a two-year appointment," Martin informed Oakley after some small talk. "And it's rare that we offer a tenure-track position after its completion."

The conversation was not turning out well. Oakley decided to forego the introduction to the subject and bring up the obvious. "What you're saying, Ted, is you don't approve of my gambling experiments."

"Ed. I'm not the type to play games, you know that," said Professor Martin. "Gambling or mind games."

"But you don't approve," Oakley continued. He wanted any pain that he was to experience to be quick.

"If I didn't approve, Ed, I'd say so. Besides, I like to think of you as another Norbert Weiner. And his eccentricities are legendary."

"I'm eccentric?" asked Oakley, beginning to be relieved.

"You're a mathematician, Ed. And eccentric is good. Professor Shannon and you get along very well, building those roulette-computing machines and testing them in Nevada. Someday they'll be in the MIT museum, I'm sure. That's good. Eccentric is good."

"But. . .," Oakley began, not giving up on his attempts at suicidally throwing himself into the fire with Giordano Bruno. He was interrupted by two students passing by.

"Hi, Professor Thorp," said the woman student.

"Hi, Judith. Hi, Flash," Oakley replied, knowing, of course, the names both of his student and his student's girlfriend.

"However," said Professor Martin, continuing Oakley's truncated sentence, "some of the other faculty don't approve." He pointed to the two students. "But the students love you," he added.

"Yeah," said Oakley, resigned to what was coming next.

"Frankly, I think it's part jealousy," said Professor Martin. Oakley was getting annoyed that Martin was taking the academic knife and churning it deeper and deeper into his abdomen instead of just relating his obvious decision. "You're, what, thirty years old?"

"In August," said Oakley.

"Twenty-nine years old. You've gotten more recognition for your work than just about anybody here," noted Professor Martin. "Not to mention the television appearances."

"Not the best publicity for the department, I admit," admitted Oakley.

"If it attracts good students, who cares?" stated Professor Martin. "But it doesn't matter to me, Ed. The point is, as you know, the instructorship is a two-year appointment."

They had reached the permit annex parking lot north of building 2 across Vassar Street and now stopped by Oakley's red T-Bird. Oakley looked at the fender of the T-Bird, wet his finger, and rubbed a spot.

"Dent," he explained to Professor Martin. This seemed to be a trick motif he used to diffuse painful situations, and Martin's last comment made it clear that this was going to be a painful situation.

Professor Martin pulled a letter out of his shirt pocket and handed it to Oakley. Oakley stopped repairing his car, took the letter from Professor Martin, and held it without reading it.

"Besides," he said, "Ralph Crouch wanted to know when you'd be available. He wants you to join his game theory group at New Mexico State. Perhaps in the fall."

Oakley laughed. "You're kidding" were the only words he could say. The thoughts and emotions that instantly welled up within him were synergistic in creating unbounded joy. Losing this position would have been a difficult event in his marriage. Vivian was on the west coast, where she was happy, her music and design careers going well. Being suddenly unemployed might have been the reason for her to remain there. Now, in contrast, he would be relocating to the west and they would both be happy, personally and professionally.

Oakley's academic career, instead of being jeopardized, was apparently flourishing. At the end of his appointment in the Moore instructorship position he would join the faculty at New Mexico State. It was obvious that his investigations, "research" perhaps being too lofty of a term, into blackjack had been not only recognized by peers in game theory but applauded. All was well.

Professor Martin, for some reason, had been toying with him. Perhaps it was some sort of intellectual game that he, of higher academic status than Oakley, wanted to play. Perhaps he was the faculty member to whom he had referred as being jealous, although Oakley doubted that, knowing that Martin did not, as he had said, play games, gambling or mind games, in his words. Perhaps he just wanted Oakley's pleasure to be that much greater, rebounding from the emotional abyss of the Mindanao Trench to the emotional pinnacle of the peak of Mt. Everest.

"They are very high on you, Edward. Ralph said your research on blackjack is the greatest achievement in probability since Cardano. He thinks you have made yourself an immortal."

Oakley chuckled briefly. His mind was overflowing with multiple stimuli, the recollections of the immortality conversation by the Truckee River, feelings of humility mixed with amusement that he would be spoken of in the same sentence with the famed sixteenth century mathematician Gerolamo Cardano, the developer of the field of probability theory, and his eagerness to read Crouch's letter.

He read the letter silently.

"Ralph also wants someone who can relate to undergraduates," Professor Martin continued. "And, it seems they have a bunch of guys that regularly visit Nevada. Ralph, you'll find, is quite the outgoing character. No doubt they're all interested in your work."

Oakley read excerpts from the letter out loud, perhaps to make sure he wasn't dreaming.

"Tenure track professorship, I get a corner window office, I only have to teach one class a semester, and I get my own university car for field trips."

"I didn't know mathematicians take field trips," he observed to Martin, catching his breath.

"They must really be interested in your work," Professor Martin replied.

*　　*　　*

Oakley was concerned about the trip to Puerto Rico with Manny Kimmel and his nieces. There was the crime problem, traveling off the mainland to, if not a foreign country then a commonwealth with a significant low-income population rather than a state. Traveling with two ravishing, voluptuous blondes to Puerto Rico seemed to be asking for trouble, even with Manny's gangster reputation protection. He wasn't a large man, and Oakley, though over six feet tall, was slight of build.

He, in addition, couldn't think of Vivian without accompanying guilty feelings. He wasn't abandoning her, but he was ignoring her feelings. Perhaps to divert his conflicting thoughts, perhaps to let Vivian know he was thinking about her, perhaps to prevent himself from making a potentially disastrous phone call, probably for all these reasons, he decided to record and then send a message. Because it would arrive at Allan's when he was in Puerto Rico, none of them could use the communication to try to dissuade him.

A tape recorder, telephone, several darts, a photograph of Vivian, and a large envelope lay on the living room table in their rented home in Brookline. A music stand propping up a piece of sheet music stood next to the table. A dart board hung on the wall.

While he held his trumpet, Oakley sang the lyrics to "These Foolish Things" into a microphone connected to the tape recorder.

> *A cigarette that bears a lipstick's traces,*
> *An airline ticket to romantic places,*
> *And still my heart has wings,*
> *These foolish things remind me of you.*
> *You came you saw you conquered me,*
> *When you did that to me,*
> *I knew somehow this had to be.*

He placed the microphone on the table, picked up his trumpet, and played four bars of music. He bobbed his head and smiled in admiration at the licks he had improvised and then placed his trumpet on the table and picked up the microphone.

He continued singing.

> *Oh, how the ghost of you clings!*
> *These foolish things remind me of you.*

Finished, he spoke into the microphone.

"I'm sure you don't know who this is," said Oakley. "Here's a hint. I'm not much of a singer, but you once said I was the greatest trumpet player in the world. Vivian, I have only one more of these gambling experiments. But I've gotta do it. It's the last one. I promised you that I'd never lie to you again and

I'm keeping that promise. I miss you very much. And I love you very much. But I gotta do this. Ed."

Oakley was sure that the song would remind Vivian of the moments that Sunday morning he had gotten onto the bus to go back to UCLA after the Pacific Coast Collegiate Jazz Festival, where they had met. He removed the tape from the tape recorder and placed it into the envelope. He stared alternately at the telephone and the darts.

He picked up the phone. He had decided to send the package rather than call her, but he began to reconsider, needing to talk to the woman he loved. He thought for a second about the possible repercussions and then quickly put the phone down. He picked up the darts and forcefully threw one toward the dart board on the wall. It missed and fell to the floor.

19

Holding Hands at Midnight

Most American gamblers have never played in Puerto Rico. The expenses of the plane flight and accommodations deter many. The very name of the U. S. territory itself scares many people based on tales of Puerto Rican gangs in major American cities. It's logical to assume that only the affluent can reach the U. S. mainland, and if their children become gang members, then those left behind in the squalor of the low-income barrios must be that much more violent. Combining that with activities in the gambling business would repel most people, especially if you expect to carry cash with you into the casinos. Manny Kimmel, as a well-known if not particularly feared American gangster, had implicit protection. You can mess with American tourists, but you don't mess with American gangsters or the street cleaners might find your body lying in some gutter. Oakley wasn't that concerned. He considered Manny as effectively being his bodyguard.

Santurce is one of the districts of San Juan, the largest as well as the capital city of Puerto Rico, located on the north shore in the eastern part of the island. Within Santurce is El Condado, an oceanfront so-called sub-district of about 3000 middle and upper-class residents. El Condado was home to several casinos, but, because it also housed an expansive resort, Manny preferred staying at La Concha over other hotel-casino locations in El Condado such as the Diamond Palace, Caribe, and Condado Plaza. He had traveled to El Condado in the past, and he knew that his nieces could entertain themselves in the resort, rather than potentially disrupting Oakley's gambling.

None of these casinos were as lavish as the casinos of South Lake Tahoe or the higher end casinos in Reno, but, as Oakley would say, it's the same game

© Springer International Publishing AG 2017
L.M. Golden, *Never Split Tens!*, DOI 10.1007/978-3-319-63486-9_19

at every table at every casino. That is true except for minor rule variations, most notably, whether the dealer has to stand on soft 17, that is, a hand composed of an Ace and a six. In Puerto Rico, the rule stated that the dealer has to stand on soft 17, just as in most Nevada casinos.

Oakley felt that he would most likely never return to these casinos and that any photographs taken of him in Nevada casinos would most likely not have been shared with their Puerto Rican counterparts. As a result, whenever the deck became favorable he was betting the unfortunately low house limits of $50. The Puerto Rican dealers dealt more quickly than those in Nevada, as Manny had mentioned, somewhat offsetting the low table limit in determining the player's potential action each hour. When they were available, he was laying bets on up to the entire seven positions on the table, further increasing the potential action.

The casinos, unlike in Nevada, were open only between 8:00 p.m. and 4:00 a.m. This necessitated a stressful roughly twelve hour adjustment in Oakley's daily schedule from his work day in Cambridge. The first four days, Oakley played in the late evenings and mornings at the Caribe, the Diamond Palace, and the Condado Plaza, hitting each casino for one to two hours on a given shift, and then taking a break and waiting for the next shift to repeat the cycle of visits. In the early evening, after dinner with Manny and his nieces, he played in the friendly confines of their home base, La Concha. This was really the only time that the girls would watch, Manny demanding, as any good solicitous uncle would, that they remain inside the resort.

This was, to tell the truth, marginally violating the paper route technique of not winning too much or staying too long at a given casino in order to avoid excessive scrutiny. Although he wasn't staying at a given casino more than a couple hours, when playing with a favorable deck he was both betting fairly heavily, near or at the table maximum of $50, and taking multiple hands. More importantly, he was winning substantially. The paper route technique would direct the player to leave a casino if doing so. Nonetheless, feeling, as I said, that he'd never come back to gamble in Puerto Rico, Oakley hoped that the casino bosses were just waiting for his luck to turn. The paper route technique was again showing its value, even if Oakley was marginally violating it.

They had a reason to hope that Oakley's luck would turn. Although early in the playing he had been down about $2000, he had, in the first four days of playing, won about $700 at the Caribe, $1200 at the Diamond Palace, $1100 at Condado Plaza, and $1400 at La Concha. That totaled about $4400 compared to the $6000 he had made playing in South Lake Tahoe and Reno. If the table limit had been $500, as it was in most of the Nevada casinos, and he were playing just one hand at a time, as he had in Nevada, instead of the four hands he had been typically playing in Puerto Rico, he would have made about twice as much.

Knowing his winnings, and now that he was keeping a log, he could get a reasonably good estimate of the power of the complete point count system in practice and could compare it to Julian's computer predictions. Oakley estimated he had played about 28 hours, so that he was making about $150 per hour, which seemed great. It was, after all, almost as much, Oakley lamented, as the head coaches at Boston College made.

He, however, wanted to calculate if these winnings were consistent with the predictions. First he had to calculate his action, how much he had bet. Figuring that he was betting a minimum of $20 at an average of four positions, his effective minimum bet per round of hands was $80. His effective maximum bet per round was similarly four times $50 or $200, so that his average bet per round was about $140. He estimated that he had played 100 rounds each hour on the average, so that his total action based on an average bet of $140 was 100 rounds per hour times 28 hours times $140 per round, or about $392,000. Winning $4,400 meant his rate of winning was 1.1%, nearly ten times the rate of winning of 0.12% with the basic strategy.

Although that was a respectable rate, the highly advantageous very "hot" deck promising dazzlingly large winnings hadn't yet appeared. Julian's calculations had shown that after playing roughly half way through the deck, limiting his maximum bet to only twice his minimum bet, he should have been making about 1.5% of his action.

Another reason for continuing existed. His total winnings had basically doubled Manny's bankroll, which was the so-called business agreement. Manny, though, to seal the business deal, had demanded that he make that $10,000 as a profit, after expenses. The expensive little vacation to La Condado required Oakley to play on.

The winnings at the Caribe would have been greater, Oakley rationalized, but he had been cheated. The dealer seemed extraordinarily clumsy. On repeated hands, two cards would keep trying to come through the slit in the shoe and would get jammed, followed by the dealer's fumbling around to extract them. When Oakley moved to another table, the shoe was moved with him. When Oakley moved to a third table, the shoe was remarkably moved with him again. It seemed as if the shoe had been manufactured to allow the dealing of seconds, as Manny had described at the consultant session in Brookline. Either the casino had become aware of Oakley through some international grapevine or American news outlet, or they simply didn't want him to take any more of their money. Whatever the case, Oakley would no longer gamble at the Caribe.

Oakley was taking advantage of two aspects of the Puerto Rican rules, having the cards dealt face up to the players and using a double deck of two fifty-two card decks, a total of 104 cards.

Having the players' cards dealt face up was a major advantage. If he were playing all seven positions, he could have fifteen cards in his count, including the dealer's up-card, aiding his strategy decisions even on the first hand he was playing. It was natural for him to stare at his own cards, and as they were dealt he could obtain the point count at the start of the hand. He could do this quickly, using the holistic technique of visually canceling out images pair wise that he had developed in Reno rather than sequentially adding plus and minus ones. In this way, no need existed to scan the table as other players busted their hands. He could easily keep the running count as each additional card was played, providing him with current information from which to make his strategy decisions.

The use of double decks also proved advantageous. He was at times using the "strike when the deck is hot" technique. He would observe a table where only one or two players sat, wait until the point count was positive, and then sit down and bet the maximum $50 bet at all the remaining places. With a double deck, the favorable decks would stay favorable longer.

He also discovered the seemingly miraculous effects of employing what he was now calling "end play." He had become aware of its possible lucrative benefits ever since Manny had told him of the rules variations in Puerto Rican casinos. The casinos would deal down to the second to last card in the deck. This enabled him to achieve a great advantage if the last cards in the deck were predominantly either low point value cards or high point value cards. If, for example, the point count indicated that the last five or ten cards were all twos through sixes, he would take multiple hands and, knowing the deck favored the house, bet low. He would draw all of those cards. When the dealer was forced to shuffle, those cards would be absent from the newly shuffled deck. Oakley would be at an advantage for the entire run of that new double deck.

If, say, ten or fifteen cards were left and most but not all were low-value cards, for another example, he would again take multiple hands and bet low. If he got a pair of tens or an 18 or 19, he would take a hit to bust the hand. He would do the same if given a natural, that is, a ten-value card and an Ace, if the casino allowed. He would draw enough cards to the hands composed of low-value cards but without busting them, to utilize all the remaining cards. The newly shuffled deck would then be replenished with the ten-value cards and Aces from the hands he had busted, but would be depleted in the low-value cards that remained on the table, as in the first example.

If, on the other hand, the last five or ten cards were all ten-value cards or Aces, Oakley would place large bets, the $50 maximum, on as many places as needed to ensure that the dealer would not have a high-value card, that is, a ten-value card or an Ace, left in the deck with which to finish the hand. The deck depleted, the dealer would have to shuffle the deck, and the card he or she would then deal to himself or herself would be from a deck depleted in ten-value cards and Aces. Oakley knew that his advantages in such end play cases could be far greater than 10%.

After all was said and done, however, it was clear that Julian's strategy modifications of the complete point count system for the Puerto Rican game were effective. The system was working as well as it had in Nevada. Oakley was content.

On his fifth day in El Condado, following the sumptuous dinner at La Concha with Manny and his nieces to which they had become accustomed, Oakley played alone at the Diamond Palace. For two hours, the decks swung between being favorable and unfavorable. When it was unfavorable, Oakley reduced his bet, but lost moderately. When it was favorable, he increased his bet, but continued to lose. After about what seemed 200 years of such anguish, a favorable deck rewarded his faith and he picked up $150 in ten minutes. He was about even.

Oakley was tired and still sluggish from the large meal and decided to obey the paper route technique and leave the Diamond Palace. Manny suggested they return to La Concha, where Oakley had had his best success, for a light snack break. He could also watch over his nieces.

Perhaps the best-known casino to the American gamblers who come to Puerto Rico is La Concha. La Concha was more finely decorated than the Diamond Palace, Caribe, or Condado Plaza but still somewhat different from the lavish American casinos. Although live palm trees were planted outside the entrance and placed in large planters in the lobby and dining areas, the only palm trees within the casino itself were plastic. Without any windows, the norm for casinos, where management tries to prevent players from knowing the time, any live plants would soon wither. The plastic palm trees in the casino showed lack of care, their green color muted by dust.

The blackjack tables were covered with the customary felt, and the design, with the betting squares, cash drop box, and chip tray, would be familiar to American gamblers. The felt at most of the tables had cigarette burns and drink stains. The dealers, almost all women, advertised their second occupation by wearing low cut white frilly blouses, showing enough curves to distract the players. Being a resort, La Concha offered numerous opportunities and locations for diversion.

Juanita, in her mid 20s, was no exception. Her name tag read "Juanita," and the sheer fabric of her blouse sagged under its otherwise inconsequential weight. Oakley had played at La Concha but hadn't yet met Juanita. He would remember her well. It was nearly midnight in Puerto Rico.

Oakley sat alone, dressed in a suit coat, tie, and long-sleeved white shirt, the required dress for Puerto Rican casinos. The shirt alone showed he was an American tourist. The 90 degree temperatures called for short-sleeved shirts.

La Concha continued to be kind to Oakley. He had bought $500 in $25 chips from the cashier, and next to that bank on the table sat his winnings, a pile of about thirteen $25 chips and ten $5 chips, a total of $350. He was playing three hands and betting a minimum bet of $25. A Jack of Diamonds and a six of Hearts were face-up on the table in front of him, next to eight $5 chips, a $40 bet. His other hands totaled 17 and 19, both standing totals against Juanita's up-card of a five of Diamonds. He had $40 bets sitting next to them as well.

Juanita busted, drawing a ten of Clubs to go with her now revealed down card of a King of Hearts. Juanita took twenty-four $5 chips from her tray and placed them, eight at a time, next to Oakley's bets. He was now up more than $450 at Juanita's table.

Juanita spoke with a thick accent. "You be lucky player, mister," she smiled.

Over the next half an hour he picked up another $125, an amount about equal to the winnings on that one hand. His initial $500 bank had more than doubled, twenty-one $25 chips and fourteen $5 chips, a total of $595, lying on the table in front of him next to the original $500 bank. He had made nearly $600 since sitting at Juanita's table. Manny was standing behind Oakley, not saying anything. He seemed content to simply count up the number of chips accumulating in front of his professor colleague.

Oakley's tie was loosened at the collar. By this time, a small crowd, attracted by the large piles of $5 chips and $25 chips piled up in front of Oakley, had assembled behind his table, joining Manny in watching the American play.

A pit boss had come by briefly, but Juanita had said some words to him and he quickly left. Juanita felt she could make more money doing a trick with this gringo, and the pit boss knew he would get a share. Such commissions, so to speak, especially when dealing with a wealthy American, must have been much more than the pit boss could earn in any San Juan casino. Perhaps the men behind the one-way windows in the ceiling were also in the consortium.

At some time during the next forty-five minutes, the deck must have communicated with that at the Diamond Palace, which had tortured him. It decided to run neutral to cold. Oakley had lowered his bet to his minimum $25 and had kept his losses down to only a bit more than $100. He was still up nearly $500. Some of the crowd had dispersed, and at one point another player sat down.

In American casinos, the dealer would have taken a break to be replaced by another. When Juanita's time had come for a break, the pit boss walked over with another woman, but Juanita said some words in Spanish to them and they left. It was one of the few times in his life that Oakley wished he had studied Spanish in high school instead of French. Soon after that brief episode the deck turned in Oakley's favor. Perhaps it was the French goddess of language showing compassion. She, it would turn out, was an extremely compassionate goddess.

Oakley was playing seven hands, each with the maximum $50 bet, two $25 chips. He had used end play on the previous deck, and the newly shuffled deck began with a point count of +5. Oakley's suit coat was draped over the back of his seat. His shirt sleeves were rolled up and his tie was hanging from his neck, unknotted. This was big business.

They played two rounds, Oakley taking seven hands in each, and the deck became more favorable to Oakley, with three more twos through sixes being played than ten-value cards and Aces. The point count was now +8. With 46 cards having been dealt and viewed, only 58 cards remained from the 104-card double deck. The high-low index was $+8/58 \times 100$, or about +13.

The complete point count system directed a bet of one-half the high-low index times his minimum bet. That minimum bet was $25, so that his bet with a high-low index of +13 should now have been $1/2 \times 13 \times \$25$, or about $150 per hand. Unfortunately, the house limit was $50. He was relegated to playing on with that bet, satisfied with taking multiple hands.

On the next hand, Juanita dealt herself an up-card of a six of Hearts, increasing the point count to +9. The cards dealt to Oakley were, as always in Puerto Rico, dealt face up. His first hand consisted of a nine of Clubs and a Jack of Spades, a pat hand. The second hand was the Ace of Diamonds and the Ace of Clubs, a hand to be split. The third two cards were a ten of Hearts and a Queen of Diamonds.

His last four hands were a mix, a six of Spades and nine of Diamonds, a ten of Clubs and a six of Diamonds, a pair of sevens, and the second six of Hearts in the deck with a five of Diamonds. With his eyes, he had quickly canceled out the various pairs of ten-value cards and Aces with the low-value cards in his seven hands and Juanita's up-card to determine that the point count had increased by +2 to +6. The high-low index, with 61 cards played and viewed from the 104-card deck was $+6/43 \times 100$, or about +14.

Oakley was in a great place. He cracked a slight grin, sensing that the action he had awaited all evening was about to begin. The small crowd was hushed, but electrified with anticipation as to what the gringo would do.

Oakley waved his hand over his first pair of cards, a pat hand. He then moved the Ace of Clubs to the right of the Ace of Diamonds. As expected from any player, he was splitting the Aces. He placed two $25 chips next to the Ace of Clubs.

"Split," he said.

Juanita dealt an eight of Clubs next to the Ace of Clubs and a Jack of Hearts next to the Ace of Diamonds, giving Oakley two more pat hands and a total of $400 on the table. The point count was +5 and the high-low index was +5/41 × 100, or about +12. She pointed to the two hands.

"Stand," said Oakley.

Oakley now played his third hand, moving the ten of Hearts to the right of the Queen of Diamonds. He placed two $25 chips next to the ten of Hearts. He was going to split the tens.

"Split," said the monosyllabic hero. Oakley's table for splitting said to split tens against the dealer's 6 if the high-low index was above +7. It was +12. Oakley had seen the interchange between Juanita and the pit boss. He was sure they didn't have his picture in Puerto Rico, and he hadn't been playing long enough to have revealed that he was playing a viable system. In any case, he reasoned, they probably had never seen his type of play previously.

The small crowd gasped audibly at his strategy choice. It was almost certain that none of them had ever seen a man split tens, an almost sure winning hand. The pit boss came over, more to see the cause of the commotion than to cause trouble. Juanita gave him a quick warning look, then turned back to Oakley.

"You split the tens, mister?" she said, her Spanish accent being even more pronounced as she lost some composure. "Never you should to split the tens," she said maternally. Oakley was touched by her affection, but then thought that her comment may have been motivated a little by greed. The more money he made at the tables, the more money she might expect to get as a tip or, better, earn on her anticipated second job. He now had $450 on the table.

Juanita dealt a nine of Hearts next to the ten of Hearts and a Jack of Clubs next to the Queen of Diamonds, two more pat hands. Oakley stared at the cards for a second. The point count had been reduced to +4, and the high-low index was +4/39 × 100, or about +10, still greater than the +7 for splitting a pair of tens against a dealer's 6. Juanita pointed to the two hands. Oakley kept on thinking. It was obvious that he had reached a point in the deck that called for some mental calculation, and he wanted to be sure that the cash on the table, the crowd, and the smiling Juanita had not caused some left-mode neurons to misfire.

Many of those who had left when Oakley ran into his cold streak had returned. Others had joined them, making a crowd of about twenty-five people standing behind the blackjack table. He waved his hand over the nine of Diamonds and ten of Hearts, but then moved the Queen of Diamonds to the right of the Jack of Clubs.

"Split," he said, announcing his decision as he put two more $25 chips next to the Queen of Diamonds. Five hundred dollars of Oakley's money now lay on the table. He had only laid down more at Harold's when he doubled and split to have $900 on the table, but the table limit there had been $500 and here it was only $50.

Juanita placed the second Ace of Diamonds in the deck next to the Queen of Diamonds, a nice touch, Oakley thought, and a King of Spades next to the Jack of Clubs. He wondered if Juanita had seen all the Diamonds on the table and had begun to dream of baubles that this American might bestow in his largesse upon her. The count was +3 and the high-low index was now +3/37 × 100, or about +8. The splitting table of the complete point count system directed Oakley to split the tens, for the third time, saying split tens against the dealer's 6 if the high-low index was, as I said, greater than +7. This was clearly getting into murky waters. Although the pit boss was not interfering with Juanita's seduction efforts, Oakley had no idea if he was being observed from above. Splitting again was sure to generate attention, at least from the audience standing behind him. The index was +8. The system he was using was his.

He moved the King of Spades to the right of the Jack of Clubs and placed two $25 chips next to it. That's $550 and counting.

"Split," he announced over the casino loudspeaker. It might as well have been the loudspeaker.

He didn't notice the many comments from the fans because something was pushing against his back. He turned to see Manny frantically shaking his head "No." Oakley had never previously seen a man kill himself by breaking his own neck.

Juanita saw this and might have figured this man was Oakley's father.

"Mister, I give to you the chance no to split the tens." Oakley believed she was concerned for him. Maybe she was thinking this man was *loco* altogether. Oakley shook his head in rejecting the offer and Juanita dealt. She placed a seven of Spades next to the King of Spades and a ten of Diamonds next to the Jack of Clubs. A couple voices in the crowd began a subdued chant of "split . . . split . . . split." Spanish voices commanded, "*divida . . . divida . . . divida!*"

The point count was now +2 and the high-low index was +2/35 × 100, or about +6. He stood with the seven of Spades and the King of Spades. Juanita pointed to the second split hand, the Jack of Clubs and ten of Diamonds. With the high-low index at +6, however, no more splitting was called for. Oakley's disappointed fans would have to wait by the players' entrance to get his explanation as well as autograph.

"Stand," he said. This caused both confusion and relief among those watching. This guy had already split a pair of tens three times, some may have thought, so he should do the same again. It's a stupid play, usually, they also may have thought, so maybe it's good he finally just stood. Juanita took a deep breath, perhaps in relief that her man would not place any additional foolish bets. Manny, his head apparently still connected to the rest of his torso, removed the pressure on Oakley's back.

Following this brilliant assault, Oakley arched his head back to crack his neck vertebrae to relieve some tension. He still had four hands to play. Juanita pointed to his fourth hand, that of a six of Spades and nine of Diamonds. Oakley's hard drawing and standing table said to stand with a hard total of 15 against the dealer's 6 if the high-low index exceeds −28. As Oakley waved over the cards to say he was standing, some murmurs were heard in the crowd. He similarly stood on the fifth hand, the ten of Clubs and a six of Diamonds. Now he faced a pair of sevens. The gamblers' saying goes "always split Aces and eights," with other pairs to be split according to the pain tolerance of the player.

Even Oakley's basic strategy, however, directed the player to always split sevens against the dealer's six, a strategy in common with the complete point count system.

"Split," said Oakley, for the fifth time in the last minute and a half. He had split Aces, tens, and sevens. One might have thought he was auditioning for a job as a ballet instructor. He placed another $50 next to the first seven. The table held $600 of Oakley's chips.

Juanita sighed, lowered her head in deep sorrow, but obliged, delivering a Jack of Diamonds and a five of Spades. The point count was now +2 and the high-low index was +2/33 × 100, or about +6. His decision about the first hand, a total of seventeen, was easy. He waved his hand over the cards to signify he was standing.

The second split hand, a total of 12 against the dealer's 6, is one of those that cause consternation among blackjack players. It is as close to a "call it" as exists. The hard drawing and standing table directs the player to stand if the high-low index is greater than zero. In other words, half the time you should stand and half the time you should take a card. No wonder it causes the consternation among players that it does.

Indeed, of all the entries in Oakley's five strategy tables, splitting, hard and soft doubling down, and hard and soft drawing and standing, only five entries were zero. These were splitting fours against the dealer's up-card of 5, splitting sixes against the dealer's 2, hard doubling holding 9 against the dealer's 3, soft doubling 19, that is, the Ace and eight, against the dealer's 6, and, alone among the entries in the hard drawing and standing table, Oakley's presently held 12 against the dealer's 6.

Juanita was pointing at the hand as Oakley remained in his reverie. A quantitative means of making these borderline decisions was a major strength of his systems. It was one of their *raisons d'etre*. Juanita had no idea why Oakley uttered his next statement with such authority.

"Stand," he said.

As a result, the high-low index remained +6 as Oakley doubled down with his final hand, placing two more $25 chips next to the six of Hearts and five of Diamonds. The hard doubling table directs the player to double down with an 11 against the dealer's 6 if the index exceeds −35. Under Puerto Rican rules the player can only double down with an 11.

He had $650 on the table. She dealt him a five of Clubs, for a lousy total of sixteen. The rules of blackjack state that if you double down you get one and only one card. Even if he could have taken a hit, he wouldn't have with the large value of the high-low index and the dealer showing a six. Oakley would be happy if only that hand was a loser. No, he'd be ecstatic.

Oakley having played his seven hands, which had become twelve, it was now Juanita's turn. The point count was +3 and the index was +3/32 × 100, or about +9. With such a preponderance of ten-value cards left in the deck, Oakley hoped that Juanita would bust.

It's good that Oakley was seated or else the crowd squeezing to get a look at Juanita play would have crushed him and his loosened tie against the table, perhaps tilting it and spilling those flat multi-colored cardboard wafers in the chip tray that he was so politely trying to extricate, mathematically speaking.

Juanita flipped over her bottom card to reveal a King of Clubs, giving her a total of sixteen. Less than seventeen meant she had to take another card. She stared at Oakley. The fans stared at the table. Oakley stared at her hand approaching the shoe. She dealt herself a nine of Spades.

Her smile this time was not subtle as the crowd cheered and applauded. She seemed genuinely happy for Oakley.

"Juanita bust. Juanita bust for you. You like Juanita bust?" she asked, pointing to her name tag and the possible treasures over which it rode. He was now up more than $1000 at La Concha. In Nevada, with a table limit of $500 as opposed to the $50 table limit in Puerto Rico, that $650 win could have been

$6500. He thought of Vivian. He wondered if she were here, would she have seen smoke coming out of his ears.

Worried about possible repercussions upstairs, he decided to bet only four places on the next hand. That bit of camouflage might indicate he was lucky, irrational, and in fact not system-addicted. He placed four bets of two $25 chips. Juanita dealt two cards face-up to Oakley at each position and one card face-up and one card face down to herself. The cards to Oakley uncharacteristically fell short of the betting positions, hitting the table between Oakley and Juanita. Oakley reached out with his hand to pull his cards toward him. Juanita placed her hands on his hand.

She spoke to him, softly, breathing deeply, as if being strangled by genuine passion. It was time for the old closing pitch.

"You very good player, mister," she said. "You good at more?" Maybe tonight you show to Juanita, huh?" Juanita, it appeared, was a very affectionate young lady.

Oakley thought Juanita deserved a break. She had been dealing to him non-stop for three hours. Normally, a shift change occurs every half an hour or forty minutes, but Juanita's understanding with the pit boss had allowed her to continue dealing.

"Tennis. I'm also good at tennis," Oakley said, scooping up his chips. Normally if you bet and get cards you have to play. Either the pit boss wasn't attentive or he was glad this gringo wasn't going to take more money from his pit. Juanita, still hoping for additional income, didn't complain.

Oakley got up and walked over to the cashier, with Manny closely beside him, protecting Oakley from potential evil-doers, like an armed guard on horseback riding alongside the Wells Fargo wagon of the Old West. Madeleine and Renee had been playing roulette, but, seeing Manny and Oakley at Fort Knox, they walked over to them.

One can only wonder what Juanita thought. Here was her ticket, walking away with an older man and two healthy, well-fed, blonde American women. Perhaps she thought Manny was Oakley's father and the girls were somehow related. More likely, she would have guessed the girls were in the same occupation as herself. Then, she must have thought, what purpose did the older man serve. The forlorn Juanita was confused.

She stood alone at her table in the background as Oakley, Manny, Madeleine, and Renee walked past banks of slot machines. Manny held the stack of cash.

"Nice," he said smelling nearly $1700 in various denominations of dollar bills. "See Thorp, stick with a sharpie like me and you can get rich." His bankroll of $10,000 had turned into $21,000. He was quite content.

"No," countered Oakley, "stick with a mathematician like me, and you can get rich." The system had proven itself, in Nevada and now, with a different set of rules, in Puerto Rico.

Manny was happy and laughed. "Well, we're gonna have a party tonight, right gals?"

"Professor Thorp, you going to party with us?" asked Madeleine. Oakley, tall, lanky, professorial, wasn't exactly a symbol of testosterone gone berserk, but that winning system might make someone even as physically gifted as Madeleine reconsider.

"I'd like to," said our party animal, "but I've got some lectures to prepare."

"I wasn't too good in school," said Renee, stating what was no doubt a revelation to Oakley. "But, I'm sure you'd be a very, very good teacher." She wasn't going to let Madeleine get ahead of her in this chase. Manny was a good provider, but he didn't have time on his side.

"What are the lectures about?" asked an intrigued Madeleine.

"Degeneracy," answered Oakley. He was telling the truth, but was also interested in the possible responses of the girls.

"Oh, wow," blurted out Renee.

"Degeneracy in white dwarfs," said Oakley, again telling the truth and again anticipating the response of the girls.

Madeleine delivered. "That sounds so kinky."

"Well, the dwarfs like it," said Oakley, enjoying the repartee with two gals who didn't know what he was talking about. He made a mental note to tell them about Subrahmanyan Chandrasekhar's research on the plane home. "It basically gives them eternal life."

Renee's head jerked back in puzzlement as Madeleine uttered, "What?" Manny's nieces were now probably as confused as Juanita.

The multi-layered conversation ended as they walked past a bank of phone booths. It was nearly 11:00 p.m. in Puerto Rico, about 8:00 p.m. in the San Francisco Bay area.

"Gotta make a phone call," Oakley informed Manny. He walked over to the phone booths and entered one. He dialed "O" for the operator at La Concha and asked her to make a call to the United States for him. The call would be billed to his room, with a small surcharge. He could not have cared if they got a couple dollars back from the hundreds he had just won.

He waited a few seconds while the operator placed the call. He was calling Allan. He had obtained the phone number of J and V's Design Shoppe, but at this moment he was too happy to allow any possibility of a negative response from Vivian to sour his mood. Besides, the tactical warfare plan he had been formulating for the last several months called for a surprise attack.

"Allan," Oakley said to Allan's "hello." "This is the guy who just broke the bank at the La Concha Casino in Puerto Rico." He paused to hear Allan's shocked reply. "Right. Puerto R-r-r-rico," Oakley said, trilling the initial 'R.' "It works. Even putting in Julian's modified coding for the Puerto Rican rule variations. Nobody can doubt it, Allan. I've been told that this work may make me immortal."

Oakley listened to Allan's happy response and laughed. "Yeah. Look," he said. "Don't tell Vivian or Joan, but we're flying back tomorrow and then Friday I'm coming into SFO."

Allan must have been happy to hear that his friend and colleague was coming back, the experiment a success. He commented about the stealthy surprise being planned for Vivian.

Oakley laughed. "I know. It's TWA one thirty-seven. Got that? One-three-seven. Yeah, I know it's the magic number. Just serendipitous. ETA about seven forty-five your time. Pick me up, okay?" Allan agreed.

Oakley faced a lot of relationship mending and wanted to make sure Allan understood. "Tell Vivian the experiment is over, but don't tell her I'm coming in, or you'll blow the whole thing. Okay?"

20

A Little Jazz Music

At the very west end of Golden Gate Park in San Francisco, sitting above the boulders placed to retard wave erosion, sits an establishment that should be in the dictionary under the definition of "iconic." Dick's At The Beach is a misnomer, there being no bathing in the cold Pacific waters off of San Francisco, not to mention no sand, just the boulders.

The jazz club, small even by jazz club standards, is as far away as you can get from the hubbub of San Francisco and still be in San Francisco. A small stage, at the east side of the venue, looks out over perhaps ten circular tables illuminated by those ubiquitous candles in a white plastic-mesh covered glass globe. Two or three chairs stood around each table. The musicians standing on the stage can look out the west-facing window to the Pacific. Although it was usually difficult to see stars at low elevation through the mist of the ocean, if the Moon was in the right place, the view was more intoxicating than most of the wines served, and spiritual if you were so inclined to that San Francisco sensibility.

Dimly lit, Dick's At The Beach couldn't have been more conducive to lovers if the powerful Alioto family had placed a miniature Trevi Fountain by the window facing the ocean. Oakley had wondered why the proprietors, maybe Dick, hadn't come up with the marketing phrase, "The Jazz Club by the Ocean in the City by the Bay." He thought it was pretty enticing.

Nearly all the tables were filled with patrons this Friday night. A piano and piano bench, drum set, acoustic bass, and a stool sat on the stage with microphones strategically placed in front of them. A music stand stood next to the bass and a thick book of music in a dark blue three-ring binder sat on the

© Springer International Publishing AG 2017
L.M. Golden, *Never Split Tens!*, DOI 10.1007/978-3-319-63486-9_20

piano. A similar book lay on the floor next to the bass. Stage lights hanging from the ceiling illuminated the stage.

Vivian was standing next to the stool singing "Teach Me Tonight," accompanied by a drummer, bassist, and pianist Andy Narell.

Starting with the A B C of it,
Right down to the X Y Z of it,
Help me solve the mystery of it.
Teach me tonight.

Wanting to avoid being spotted by the chanteuse, Oakley, Allan, and Joan had waited by the door next to the stage to enter. They were led to a table near the window, obscured by the darkness in the club. Vivian, partially blinded by a stage light hanging just in front of her, could see three patrons being led by the hostess, but couldn't distinguish their features. She did note that one of them was carrying some sort of small suitcase or overnight bag. It must have been a tourist just in from SFO she thought, flattered that they would come to see her before even going to their hotel. It had been a gig lasting several months and her reputation must have been growing; she congratulated herself.

Oakley carried a trumpet case, not an overnight bag. The three sat at the table, silhouetted by a rising crescent Moon. From the stage, you could see the shimmering of the moonlight on the Pacific Ocean.

Vivian continued to sing as a cocktail waitress approached Oakley, Allan, and Joan's table.

"Hi, I'm Caralinda. What'll it be, guys?" Caralinda asked.

"A pitcher of Coors?" Oakley asked Allan and Joan.

"Sounds good," said Allan.

"Excellent choice," said Joan, as if Oakley had picked a 1933 French chardonnay from the vineyards outside Nancy.

Oakley reached into his pocket and pulled out a $5 bill.

"Caralinda," he said to the waitress. "Does the singer take requests?"

"Sure."

"Ask her to sing 'I've Never Been in Love Before,' " said Oakley, handing her the $5 bill.

"Sure thing. Great tune," agreed Caralinda.

"Thanks," said Oakley as the waitress left to deliver the goods. Vivian continued her song.

The sky's a blackboard high above you,
And if a shooting star goes by,

I'll use that star to write 'I love you,'
A thousand times across the sky.
One thing isn't very clear my love,
Teachers shouldn't stand so near my love.
Graduation's almost here my love.
You'd better teach me tonight.

The chorus completed, Vivian walked over to the piano and stood next to it as Andy took a piano solo. Then she walked back to the microphone and continued the tune.

One thing isn't very clear my love,
Teachers shouldn't stand so near my love.
Graduation's almost here my love.
You'd better teach me tonight, tonight.
What I need most is post graduate
What I feel is hard to articulate
If you want me to matriculate
You'd better teach me tonight.
Come on, baby.
Come on, honey,
You'd better teach me tonight!

Vivian tilted her head slightly toward the patrons, as they applauded and whistled at her styling, voice, and witty original final lyrics.

"Why thank you so much," she said graciously, smiling. "Thank you."

Caralinda stood in front of the stage and spoke briefly to Vivian. Dick's At The Beach was an informal, laid-back, in the San Francisco vernacular, club. Request for songs were common, and Vivian, with a large repertoire and an able back-up band, usually could deliver.

"Oh, gosh! Oh, gosh!" Vivian said, visibly moved. She walked over to Andy and spoke to him.

"Jon," said Andy to the bass player, "three fourteen in the blue book."

The bass player picked up the book of music from the floor and put it on his empty music stand. He flipped some pages.

"We've gotten a request for a song that is very, very dear to me," said Vivian. "And since Andy Narell knows it, we're going to do it!"

Joan leaned over and gave Oakley a kiss on the cheek. This was very cool, she had to admit. She didn't think Oakley had this much romance inside his brilliant body. Allan slapped him on the back. The plan was working as well as the complete point count system had worked.

Vivian turned to the trio and began to count off the tune, "One . . . two . . . one two three," and she sang:

I've never been in love before,
Now all at once it's you,
It's you forever more.

Oakley pulled his trumpet out of his trumpet case and stood up at the back of the club by the window, framed by the moonlight flickering on the ocean. He began to play jazz licks behind Vivian's singing.

I've never been in love before,
I thought my heart was safe,
I thought I knew the score.
But this is wine that's all too
Strange and strong.

Vivian heard what sounded like the trumpet stylings of her husband. She placed her left hand over her eyes to shield them from the stage lights and stared out toward the trumpet sounds, bobbing her head back and forth, to try to determine if it indeed was Oakley.

The puzzled members of the trio exchanged questioning looks at this strange behavior from Vivian as well as the unannounced trumpet sounds coming from the moonlit window table. She finished the tune, with Oakley interjecting jazz licks. She certainly recognized them as his.

I'm full of foolish song,
And out my song must pour.
So please forgive this helpless haze I'm in
I've really never been
In love before.

The patrons applauded and whistled and looked back and forth from Oakley to Vivian and back. Something unexpected was going on this Friday night. Something obviously sweet and touching, as expected at Dick's At The Beach.

Mom put down her microphone and yelled "Edward!" clearly audible despite the audience applause. She ran off the stage dodging tables and chairs through the audience toward my dad and, not severely injured by the multiple collisions, embraced him. The patrons decided the romantic drama, in the moonlight and candles at Dick's At The Beach, was worth a standing ovation.

Vivian cried, "Oh, Edward! Edward!"

ALMOST THE END

21

A Milk-Bone for Gauss

November, 1967
Thorp Home
Irvine, California

Three place settings lay on the morning kitchen table in the Thorp home outside Irvine, California. Oakley had enjoyed working with Ralph Crouch and his group at New Mexico State, but the University of California at Irvine had made him a generous offer. He would spend the rest of his academic career in their mathematics department. Later he would receive a joint appointment in finance. I'm, however, getting a little ahead here.

Vivian was happy, returning to her native California. Oakley was pleased both at his academic succession and at his being closer to the casino action. A triennial gambling conference had begun in Las Vegas, and by now Oakley had become a celebrity, not only there but in the entire casino world. The first, 1962, edition of *Beat the Dealer* had made *The New York Times* best-seller list. The 1966 edition incorporated the improvements made by Julian Braun using his powerful IBM 7044, to Oakley's pleasure. Oakley had repeatedly put gracious acknowledgements of the contributions of both Julian and Harvey Dubner in the new edition. He even talked about Allan Wilson. He credited the Four Horsemen of Aberdeen and his professors for their contributions.

Vivian had set her normal wholesome breakfast meals on the table. At one place setting lay a bowl of cereal and a glass of milk. That was for Fredric, about three years old, my older brother. At a second lay pancakes, a cup of coffee, and a doughnut. That was for Oakley, who was, as his habit, reading the newspaper before driving over to UCI.

© Springer International Publishing AG 2017
L.M. Golden, *Never Split Tens!*, DOI 10.1007/978-3-319-63486-9_21

Their dog Gauss lay next to the table, waiting for whatever treats might be offered. He rarely got any other than doggie treats, the Oakley's being ahead of their time in not imposing human food with its additives, sugar, and other dangerous ingredients on canines.

Vivian placed glasses of orange juice before Fredric, Oakley, and at the third place setting, hers.

"I squeezed fresh orange juice," she said. "You'll both need the energy."

"Julian's company is doing well," noted Oakley.

"And I'll compost the peel and pulp for the Earth. What's it at?"

"Forty-seven dollars," said Oakley.

"Just last week it was forty-two," recalled Vivian. She turned to Fredric and gave him a dog biscuit.

"Honey, here's a Milk-Bone for Gauss. You know," she said to Oakley, "you should call him and say 'hello' sometime."

"Yeah," said Oakley. "Forty-seven and a quarter, actually." Suddenly his entire attention was directed to the stock prices. He was beginning to go into one of his trances. He dipped his doughnut into his coffee and took a bite.

"DuPont is at twenty-four," he said. He paused. "And a half."

Seeing his dad dip his doughnut, Fredric dipped the Milk-Bone into his glass of milk.

"Honey, don't dip it in the milk," said Vivian to Fredric, chuckling. "What's my Sears doing?" she asked Oakley.

Fredric pointed to the Milk-Bone box. "Milk bone!" he said emphatically, making the apparent liquid connection before lowering it to Gauss.

Oakley laughed at the indisputable logic and kissed Fredric on the cheek.

"Thirty-three," he replied to Vivian. "Even. Kelloggs is twenty-eight and a half." Oakley was finding magic in the fractional prices.

"Daddy, does me have a Sears?" asked Fredric. If mommy has a Sears, then maybe daddy has a Sears, and maybe Fredric has a Sears, he reasoned.

"Revlon?" asked Vivian. Some of her boutique friends had been extolling their products.

"You sure do, Fredric," Oakley answered. In a sense he did, being a member of the family. Oakley did think for a moment of the consequences of Fredric being told he in fact did not have a Sears.

"Sixteen even," said Oakley, he eyes attached to the page. "Can you give me a note pad and pencil, please?" he asked her.

Vivian had seen Oakley in this detached state of mind hundreds of times in their marriage. His brain was in the mathematics mode and verbal communication became sparse.

Vivian walked over to a kitchen drawer. "It's never too early in the morning to do mathematics," she observed.

"Me, too," said Gauss. "Like father, like son" had another example in the Oakley household. In many families, the children emulate their parents, in this era usually the father as the sole wage earner. Not only do they gain respect for the particular field, but also any success on the part of the parent facilitates entry of the child into the field, even if the child is not particularly gifted in that field. Colleagues feel somewhat of an obligation to aid the entry of the children into the field of their colleague-friend. In this way, the children of actors, musicians, lawyers, doctors, writers, dentists, dancers, architects, accountants, mathematicians, and astronomers, to name just a few, disproportionately enter those fields. The uncle of one of Oakley's heroes, astrophysicist Subrahmanyan Chandrasekhar, for example, was the noted physicist C. V. Raman, a Nobel Prize winner himself.

Vivian handed both Oakley and Fredric a pencil and pad of paper. Oakley alternately looked at the newspaper and wrote on the notepad. Fredric scribbled.

"You know," he announced to Vivian, "there's a pattern here. The most common closing prices seem to be integers. Then halves, then quarters, and then eighths."

"Sixteenths?" she asked. This one of those many times that Vivian wished she had had more formal training in mathematics. She would have loved to have been able to contribute to Oakley's work. This question was reasonable, but she hadn't seen what Oakley was seeing.

"Uhh. Wait," he said, laconically immersed in the analysis. He read the paper and wrote more comments on the note pad. Then he pointed to it.

"Look, I've tabulated twenty closings. Nine are integers, six are halves, four are quarters, and one is an eighth. There may be non-randomness here," he said to Vivian, looking over his shoulder.

"That's a small sample," she said.

"The differences already are significant to within a square root deviation," he informed her. "There is almost certainly a non-randomness here. But these numbers are for the market as a whole. I have to tally the daily fractional prices for given stocks. See if there's a pattern."

Vivian sat down. "A pattern of non-randomness," she noted, beating Oakley to verbalizing the conclusion now obvious to both of them. "Like gambling games. Like blackjack."

"Blackjack! Blackjack!" exclaimed Fredric, recognizing the word that had been the topic of innumerable conversations in their household, among both family members, friends, and colleagues.

"Right," said Oakley, letting Vivian think she had deduced a concept of which he was yet unaware. "The stock market is just a gambling game," he elaborated. "The biggest gambling game. Millions of players. Hundreds of millions of dollars bet every day. Billions of dollars of action every year."

"The stock brokers are the dealers and croupiers. And the brokerage houses are the casinos," said Vivian, extending the metaphor.

Oakley laughed at Vivian's perceptiveness. "Think of that! And the stock exchanges and ticker tapes are the gambling devices. You buy a stock and you're making a bet that it'll go up. The brokers take a commission. It's just like a rake in poker, or the house percentage determined by how the rules of craps and the roulette wheel are set."

"Even the slogans are the same, 'ride it while it's hot,'" noted Vivian, again making a good point. "Or, 'don't throw good money after bad,' because either the stock is no good or the dealer is cheating you."

"Huh!" exhaled Oakley at the truth of her observations.

"'You can't go broke taking a profit,' telling you to not try to sell stock at its peak or to leave the table after winning a big pot to avoid excessive house scrutiny," added Oakley. "There are 'bull' and 'bear' markets, just like 'hot' and 'cold' streaks. You know, in the long run the market goes up so there has to be an average player's advantage. If there really is non-randomness, we can devise a system."

Oakley and Vivian paused, thinking of the consequences. Fredric continued to scribble on his notepad. Gauss had finished his Milk-Bone.

"What do you think, Fredric?" Oakley asked him.

Fredric scribbled on his note pad. Gauss barked.

THE END

Epilogue

Born on September 25, 1929, Julian Braun, was graduated from the Illinois Institute of Technology with Bachelor of Science degrees in both mathematics and physics. He obtained his Masters degree from San Diego State College (now San Diego State University) and then taught there briefly before working for Chrysler Corporation in their Missile Systems Division in Detroit. He made his blackjack mark while working at the downtown Chicago research laboratories of IBM from 1961 until his retirement in June, 1987, where he had access to the probably the most powerful computer of the era, the IBM 7044. Not only did he perform the calculations for the revised 1966 edition of *Beat the Dealer*, but also his superior algorithms led some of the major blackjack students following Thorp to enlist him to perform their calculations. After retirement, he worked as a commodities trader from his Chicago apartment. Braun never married. He died on September 4, 2000, in Chicago at the age of 70 without fanfare, without Thorp's being aware of his illness, and having been unhappy at the lack of recognition and lack of financial reward for his pivotal contribution to the theory of blackjack. He is buried in the Jewish Westlawn Cemetery, located in a near west Chicago suburb, being predeceased by his parents Marcel and Anne, his sister, and his two brothers. The author of the 1980 book, *How to Play Winning Blackjack*, he is recognized as one of the giant masters by those who have achieved renown in the field.

Dr. Allan Wilson (July 21, 1924–June 5, 2001) was in fact an acquaintance of Julian Braun at San Diego State College (now San Diego State University). After obtaining his Ph.D in physics from the University of California, Berkeley, in 1955, he was on the faculty of San Diego State College for 3 years, until 1959. (He taught there in 1954 before formally receiving his degree.) He then

© Springer International Publishing AG 2017
L.M. Golden, *Never Split Tens!*, DOI 10.1007/978-3-319-63486-9

embarked on a 31 year career at the General Dynamics Corporation facility in San Diego from 1960 until 1991. His 1965 book, *The Casino Gambler's Guide*, expanded and reissued in 1970, remains a classic. It contained a system for blackjack, the Wilson Point-Count System, that he developed and tested in casinos. The book had been in preparation for years, from the 1950's, and an earlier publication could have predated the publication of the 1962 first edition of *Beat the Dealer* and thereby rewritten blackjack history. He died in La Jolla in 2001 at the age of 76.

Emmanuel "Manny" Kimmel died in 1982 in Boca Raton at the age of 84. Founder of the Kinney Parking Company, he leased garages for storage of liquor during Prohibition. Among other gambling bookmaking activities, at one time he may have been the biggest horseracing bookie on the East Coast.

Vivian Sinetar Thorp was the owner of a design consulting firm in Newport Beach, California. She died in 2011 at the age of 82 after more than 50 years of marriage. The Thorps have three children.

Dr. Edward Thorp (born August 14, 1932) is recognized as the unequalled giant of blackjack. His 1962 book, *Beat the Dealer,* and especially the 1966 revised edition, revolutionized blackjack strategy and made it the most popular casino game. In 1967 he co-authored the book *Beat the Market,* which helped launch the derivatives concept that transformed world securities markets. Currently head of Edward O. Thorp and Associates in Newport Beach, California, his mutual fund has achieved an annual average yield of 20% over thirty years.

Of the Four Horsemen of Aberdeen, a term coined by Allan Wilson in his 1965 book, *Playing Blackjack to Win: A New Strategy for the Game of 21.* As young men of 22 to 24 years of age, they now feel that simply choosing a different title, such as *Winning Blackjack: A New Strategy for the Game of 21,* would have led to the recognition now given to them mainly by blackjack historians and scholars. The Four Horsemen became lifelong pals.

Professor Angus E. Taylor was the author of the most widely used college calculus book in the 1960s, 1970s, and 1980s, *Calculus with Analytic Geometry.* He died in 1999 at the age of 87.

Professor Robert Sorgenfrey was the individual who informed Edward Thorp of the first calculation of the basic strategy by Roger Baldwin and his associates. He died in 1995 at the age of 79.

Professor John Selfridge died in 2010 at the age of 83.

Edward Thorp, Julian Braun, and the Four Horsemen of Aberdeen, Roger Baldwin, Wilbert Cantey, Herbert Maisel, and James McDermott, are inductees in the Blackjack Hall of Fame, located in the Barona Casino near San Diego, California. Membership provides the inductee with lifetime free rooms, food and beverage service, and shows, with the proviso that the inductee will never play blackjack in the casino. The author believes that the Hall is diminished in stature by the failure of voters to have not yet elected Dr. Harvey Dubner, Jess Marcum, and Allan Wilson.

In 2003, The Edward and Vivian Thorp Foundation donated $1 million to the University of California at Irvine to fund The Edward and Vivian Thorp Chair in Mathematics.

You can get rich doing mathematics.

The Vivian Name Game

Chapter 5 introduced the Vivian Name Game. The letters of Vivian Thorp's maiden name, "Artisen" in the novel, contain seven of the eight most frequently appearing letters in the English language, lacking only the "o." Words and short content-rich phrases are produced, in the Vivian Name Game, by permuting these seven letters. Some were listed in Chapter 5. To not clutter the narrative, I provide additional such phrases here. As noted in Appendix I, Vivian's actual maiden name was Sinetar. I chose to change it to "Artisen" for reference to the creative nature of her fictional character.

In tears, when she was crowned Miss America, she was _____.

Is En Art, the painter's reply when asked about using only small dashes in his work.

Sine-Tar, related to the black magic practice of trigonometry.

Earns it, why the good doggie was given a treat after every walk.

Ne Sitar, line from the obituary of Mrs. Guitar.

Near tis, answer to the question, "Where should I look in the dictionary for the definition of 'titmouse'?"

Resinat, a product for sweaty baseball pitchers.

Tens Air, the airline formed by Bo Derek and her perfect friends.

An rites, services for the distinguished Mandarin scholar, An Hou Leong.

Ran site, why the intercontinental ballistic missile installation worker received the Employee of the Year award.

Set Iran, the announcement at center court after the doubles pair won 40-love.

Starine, a great name for a Vegas chorus line dancer.

Snare it, direction given by coach to fumbling Little League player.

Trinase, obviously a nasal decongestant, whatever that is.

© Springer International Publishing AG 2017
L.M. Golden, *Never Split Tens!*, DOI 10.1007/978-3-319-63486-9

NRA site, Belfast, Northern Ireland.

Sine Art, the style of the Renaissance painter who utilized only wiggly lines.

Stare in, psychological profile evaluation technique in which subjects ogle each other, to see who will blink first.

Retains, what is a seven-letter word using all the letters of "restain"?

Stir Ean, what Ean's mother told his father when Ean was late for shul.

Strain E, a virus for which a treatment has just been found.

En stair, an architectural technique to represent a staircase using only short dashes.

Sire Ant, the head stud of the family *Formicidae*, order *Hymenoptera*.

Sin tear, the result of a particularly stressful Catholic confession.

Nir East, the answered given by a spelling-challenged fourth grader to the question, "In what geographical region are Damascus, Baghdad, and Jerusalem."

Sire Tan, Daily Racing Form information on patrimonial lineage of MyGorgeousTan.

Rate sin, banks extorting 15% fees for loans.

Anti Res, a powerful all-purpose kitchen cleanser.

As I rent, answer to the question, "What right did you have bringing women here?"

Rain set, where they shoot umbrella commercials.

Satiner, one who makes satin.

Satie RN, title of lastest t.v. medical drama.

In Taser, command given to robot by coward cops.

Air Sent, an early non-adopted slogan of the U. S. post office.

Erin Sat, a large boulder in Ireland with mystical qualities.

Ear tins, an annoying ringing in the ear.

Restain, what the doctor told the pathology resident after he dropped the sample.

A rest-in, what the overworked garment workers called their silent protest.

Air nest, where the Floating Jellyfish People of Jupiter put their pet birds.

At rinse, the answer to the question, "Is my wife's hair done?"

Rest Ina, the chant of fans of the basketball star, Ina Hoop, after seeing she had played the whole game without a break.

Sin rate, the motion picture academy rating for pornographic films.

Sin rate, (or. ..) hooker fee.

InStare, a visual aid device for peeping Toms.

Saniter, a bathroom disinfectant.

Appendix I

Poetic Licensing

This novel does not strictly adhere to the mode of *creative non-fiction*. In it, no alteration of historical fact occurs; history is simply dramatized. Here, for the sake of the novel form, I have altered history. It is altogether fitting and proper that I declare the reasons which impel me to this alteration, the reasons for which and the manner in which I have taken liberty with history.

Their Romance

The only knowledge I have of the romance of Edward Oakley and Vivian is that they were devoted to each other. Vivian was not a student at Berkeley. In fact, they met at UCLA, she being older than him. Vivian's actual maiden name was Sinetar. I chose to change it to "Artisen" for reference to the creative nature of her fictional character. The richness of permutations of those seven letters led to the Vivian Name Game described in Chapter 5. I doubt that the Thorp children referred or refer to their father as "Oakley."

The jazz connection was invented. Neither were musicians, let alone jazz musicians. In particular, I have them throw in musical "quotes." In a jazz solo, the player sometimes throws in a few notes from a recognizable tune. Those are called "quotes." In the novel, here and there I have Thorp or Vivian throw in lyrics from well-known songs, literal as contrasted with musical quotes. Any jazz player who reads the book will appreciate these. As a graduate student at Berkeley, I was instrumental in founding the UC Jazz Ensembles and, as they

© Springer International Publishing AG 2017
L.M. Golden, *Never Split Tens!*, DOI 10.1007/978-3-319-63486-9

say, write about what you know. The Deuces of Chapter 7 and Chapter 14 was the name of the dance band founded by my twin brother Bruce and his friend Ron Svoboda in high school and "Royal Garden Blues" was one of our favorite tunes. I played in the band. In Chapter 1, I refer to his Bach Stradivarius as having an E at the top of the staff that was flat. That, of course, referred to my Bach Stradivarius. I hope the reader agrees that the jazz motif enhances their fictional romance.

As a minor note to music historians, in Chapter 8 I had Vivian refer to Thorp as "Eddie My Love" on their January 1956, honeymoon night. Although the Teen Queens recorded that tune in December 1955, it didn't reach the Top 100 singles list until the week of February 22, and didn't reach the Top 40 until the following week, at #39. It was recorded by others, including the Chordettes, whose version hit the Top 40 list that very week, at #38, ahead of that of the Teen Queens, and The Fontane Sisters, whose version hit the Top 100 that week at #74. Choice of the "Heavenly Harmony" part of the Thorps' fictitious wedding chapel was motivated by the names of the palaces at the Forbidden City in Beijing. I had the pleasure of visiting it in 1996 as a professor on Semester at Sea and photographed the signs at their entries. They, as can be verified by Chinese scholars, online photographs, or those who have visited the site, include, translated, Hall of Great Harmony, Hall of Middle Harmony, Hall of Preserving Harmony, Palace of Heavenly Purity, and Palace of Earthly Tranquility, as well as Hall of Union. "Heavenly Harmony" fit.

As an astronomer working at the Jet Propulsion Laboratory in Pasadena, I quickly learned that southern California women of pulchritude – I mean you, Carol Brooks and Sybil Abelsky – prefer lawyers and doctors. I "became" one of the former. I doubt that Edward Thorp masqueraded as a law student while courting Vivian, but it served to progress the romantic plot, embellish character, and provide some humor.

The rapid-fire events preceding the wedding of Vivian and Thorp, and the ceremony itself, were written with screenplay adoption or adaptation in mind as well as for comedy effect. The characters provide the opportunity for box-office-boosting cameo roles using established name actors at minimum cost.

The Gambling Trips and Gambling Systems

First, to avoid preliminary diagnoses of dyslexia, I should discuss briefly the results of the gambling trip, described as being to South Lake Tahoe in January, 1956, in the novel. It actually was to Las Vegas over Christmas of 1958 (see Table A.3 below), and the Thorps went alone. In the novel, Thorp won $8.50, but in fact he lost $8.50. I took that liberty so they would have money for breakfast and, jokingly, their wedding.

Concerning the later gambling trips, Manny Kimmel did bring his so-called nieces to meet Thorp in Boston, but the girls did not accompany them on their trip to Nevada. The initial trip to Nevada was financed by both Manny Kimmel and an associate of Kimmel. I opted not to clutter the narrative with the presence of that associate in Nevada. The jazz audition before Bernie Kurtzman, the namesake of a late, beloved cousin of the author's father, was fictionalized. Gussie was my paternal grandmother.

Thorp utilized the ten-count strategy, not the complete point count system, on the Nevada experiment trip. His winning $21,000 on a bankroll of $10,000 was achieved totally in Nevada.

His winning $900 on a single hand, after splitting and doubling down, occurred at a casino in Reno which Thorp did not name in *Beat the Dealer*. The table limit there was in fact $300. I changed the name of the venue in Chapter 14 to Harold's Club for the purpose of specificity. This was one of my favorite Reno casinos, not only because of its relaxed atmosphere but also because I usually almost always win there.

The trip to Puerto Rico was made as a result of Thorp's April 1964 appearance on the television program "I've Got a Secret." He was accompanied by celebrity panelist/comedian Henry Morgan and two additional bankrollers, not Kimmel. There he played, not the complete point count system but the simple point count system described in Chapter 11. Thorp did not divulge the name of the casino at which he was cheated. I used poetic license to suggest it was the Caribe. I apologize if it in fact was not the cheating casino.

Although in Chapter 11 I have Thorp and Julian Braun in 1961 refer to Roger Baldwin and his collaborators at the Aberdeen Proving Ground as the "Four Horsemen of Aberdeen," this term was apparently not used to describe them until Allan Wilson, perhaps partially being influenced by his father being an Episcopalian minister, did so in his 1965 book, *The Casino Gambler's Guide*.

Although his character does not appear in the novel, Dr. Richard A. Epstein (born March 5, 1927) must be mentioned in the context of such pre-Thorp investigators as Allan Wilson, the Four Horsemen of Aberdeen, and Harvey Dubner, the latter to be discussed in Appendix IV. As Thorp, Braun, and Wilson, he both was trained in physics, getting his Ph.D in that field, and also learned computer programming. Working as an electronics engineer for the Ramo-Wooldridge Corporation (TRW after 1958) in Los Angeles, he in the 1950's, as Wilson, the Four Horsemen, and Dubner, independently analyzed blackjack.

He calculated the basic strategy probabilities and expected winnings for the game, most likely using the Remington Rand 1103 computer available to him at work. (I have not beenable to confirm this with Dr. Epstein). He created what he referred to as decision matrices for standing and drawing, splitting, and doubling down, and calculated a slightly negative expectation for the player, similar to the original results of the Four Horsemen. Although he did not consider card counting, he discussed that keeping track of the cards that had been dealt as well as varying the bet size would give the player a net positive advantage over the house. His book, *The Theory of Gambling and Statistical Logic*, the first edition of which was published in 1967, is a classic in the field. As with Wilson's book, its earlier publication may have changed blackjack history. How many others analyzed the game in this 1950's period as computers became increasingly available in the engineering and academic world can only be surmised.

The simple and complete point count systems were not introduced until the 1966 edition of *Beat the Dealer*, 5 years after the Nevada gambling trip. I chose to have Thorp use the complete point count system for several reasons. First, it remains the most popular of the card counting systems, as described in Appendix II. Second, it requires the player to count the cards as they are played and seen. The ten-count strategy requires the player to keep track of the cards that have not been seen, that is, that remain in the deck. I find this more taxing. A player that Thorp encountered in his Puerto Rican trip was said to find it, as reported in the 1966 edition of *Beat the Dealer*, "laborious."

Third, although the ten-count strategy is comparable in power to the complete point count system, the variation in the value of the high-low index is much greater than that of the ten richness ratio and so, I believe, is more easily used by the player. The former can range to absolute values of more than 50, although most of the time the range is −30 to +30. In the ten-count strategy, the *ten richness* is defined as the ratio of the number of non-tens remaining to the number of ten-value cards. The value for a full 52-card deck is 36/16 = 2.25. Thorp provides a betting guide based on the ten-count strategy (see Table A.1). The range of the tens richness index is not only limited, but also includes one or two numbers to the right of the decimal point.

Trying both systems shows that the point count mental arithmetic is more easily performed, not requiring calculations to one or two decimal places. Table A.2 provides, for example, the values of the ten richness ratio and the high-low index for a 52-card deck of various cards having been played and seen, with 26 cards remaining in all cases, one-half of the deck. The formula for the latter, presented in Chapter 11, is $I = 100\,c/n$, where I is the high-low index, c is the point count and n is the number of cards not dealt and seen. Recall that the

Table A.1 Thorp's betting guide in the ten-count strategy	Others/tens	Bet
	>2.00	1
	2.00 to 1.75	2
	1.75 to 1.65	4
	<1.65	5

Table A.2 The ten richness ratio and the high-low index are displayed for various illustrative cases for the same number of remaining cards, 26, not yet dealt from a 52-card deck,. The table provides the number of each type of card that have been played and seen. Calculation of the former requires arithmetic to one or two decimal places. In casino play, integer values of the high-low index would be approximated by the player by mental arithmetic, and those are the corresponding results presented in the final column

2's through 6's	7,8,9's	Tens	Aces	Others/tens (bet)	H-L index by formula	H-L index (bet)
6	4	12	4	5.50 (1)	−38.46	−40 (1)
10	5	8	3	2.25 (1)	−3.85	−4 (1)
12	5	6	3	1.60 (5)	+11.54	+12 (6)
14	6	4	2	1.16 (5)	+30.77	+32 (16)
16	7	2	1	0.09 (5)	+50.00	+52 (26)

ten richness ratio refers to the cards remaining to be played whereas the high-low index refers to the cards that have been played and seen. In the complete point count system, the bet in terms of the minimum or unit bet is given as one-half of the high-low index, unless the point count is negative, in which case the bet is one unit.

As can be seen from this selective as opposed to exhaustive analysis, in the complete point count system not only is the mental arithmetic less taxing but also the betting discrimination is finer. In short, I prefer the complete point count system to the ten-count strategy, as do the vast majority of card counters. No need existed for cluttering the narrative by reporting that Thorp used the latter in his experiments in Nevada.

The term I use for the sum of the point values of the cards that have been dealt and viewed is "point count." Thorp in the 1966 revised edition of *Beat the Dealer*, in a confusing act of inexcusably slovenly scholarship, uses no less than six terms to refer to this, "point totals" (p. 76), "point-count total" (p. 76–77), "point count" (p. 78), "total points" (p. 95), "total point count" (p. 95), and "point total" (p. 96). I choose to use, as the clearest and simplest, "point count."

The various strategies for camouflage have various origins. Some were discussed by Thorp in *Beat the Dealer*. Some were invented by me in my card counting escapades. Some were conjured by both of us independently as well as by other card counters. I am proud to report that during one of my visits to the

Riverside Hotel in Reno, after its purchase by Jessie Beck and being renamed "Jessie Beck's Riverside," I had played in the morning in front of a male dealer. In the afternoon I returned and played in front of another dealer. I then took a snack break in their restaurant. The dealer who I had faced in the morning approached my booth and asked me if he could join me. He told me that in the morning he had no idea what I was doing. He then related that in the afternoon he had been watching me from the sky. He told me that, after studying my play, he concluded that I was the best counter he had ever seen. This was not so much the result of my mathematical acumen as my ability to camouflage my play.

I discuss using acting techniques as camouflage in Appendix III.

Relationships and Cameo Appearances

I created relationships other than that of Thorp and Vivian. These include those of Julian Braun and the fictitious Nancy Redman, Allan Wilson and the fictitious Joan, and the events in the Braun household. Many of the phrases spoken by Anne Braun in Chapter 2 were those of my mother. The relationship of Allan Wilson and Thorp as best friends at UCLA and gambling buddies was invented. Allan was married in August, 1954, to Bonnie Ritzenthaler as students at the University of California, Berkeley, but I opted to portray him as single for purposes of characterization. His Ph. D. was in nuclear physics, not mathematics.

Neither John Ferguson, later to name himself Stanford Wong for marketing purposes, nor the famed cosmologist Alan Guth were students of Thorp at MIT, as related in Chapters 11 and 18, respectively. Jess Marcum is reported to have developed a winning strategy for blackjack using analytical mathematics, a more difficult approach than using the approximate computer simulations method employed by Thorp, Braun, Dubner, and all subsequent investigators. He did not appear at a casino while Thorp was playing, as related in Chapter 5.

I have Manny Kimmel in Chapter 12 refer to his nephew, Seymour, during the Boston visit. Seymour Lubetkin was a mathematics graduate of the Newark College of Engineering, now the New Jersey Institute of Technology, and Kimmel claimed that he provided Seymour's finding to Thorp in Boston that the power of card counting lay in counting ten-value cards rather than fives. Thorp denies any such encounter. In the 1966 edition of *Beat the Dealer*, Thorp notes that he had begun working on the ten-count strategy even before the 1961 presentation of the five-count strategy to the American Mathematical Society. Coverage of this led to his meeting Kimmel. If Seymour Lubetkin had indeed worked on a strategy based on counting tens, it would seem that Thorp had not used that work to initiate work on his ten-count strategy.

As the reader may have surmised, the Goose and Stargazer figures, using the magic slot machine, related the saga of my twin brother and me, respectively, as we traveled throughout Nevada for an entire month one year while I was at Berkeley, as I related in the Introduction. My brother's childhood nickname was in fact "Goose." I never went by the nickname of "Stargazer." He played the ten-count strategy and I played the complete point count.

The "Hi-Lo" System, Harvey Dubner, Julian Braun, and Claude Shannon

Harvey Dubner presented the original" Hi-Lo" system at the 1963 Fall Joint Computer Conference, held November 12–14, 1963, at the Las Vegas Convention Center in Las Vegas, during a panel discussion entitled "Computers Applied to Games of Skill and Chance" moderated by Thorp and whose participants included Julian Braun and Allan Wilson. I discuss this major contribution of Dubner to the field of blackjack card counting systems in Appendix IV.

Note the title. Thorp in the 1966 revised edition of *Beat the Dealer,* referred to the title of the panel as "Using Computers in Games of Chance and Skill," and this incorrect title has been propagated through many websites and gambling books. The actual title, as noted, was provided in the pre-meeting program as published in the trade journal *Electrical Engineering* (http://ieeexplore.ieee.org/stamp/stamp.jsp?arnumber = 6,539,277). (I have not been able to ascertain if the title was changed from the date of that publication to the actual meeting, but this is not the normal practice in scholarly meetings.)

For purposes of continuity in the narrative, I implied in Chapter 11 that that system was known to Thorp before his 1961 presentation before the American Mathematical Society. Thorp then enlisted Braun to perform more detailed calculations, which were utilized in the 1966 revised edition of *Beat the Dealer.* Thorp acknowledges the contribution of both scholars.

Many today refer to the complete point count as the *high-low* system. Thorp, on page 93 of the 1966 revised edition of *Beat the Dealer,* refers to the near equality of the complete point count and and the high-low system, using the phrase "the complete point count, or high-low system. ... ." In response to my query, Thorp wrote to me, "The complete point count assigns 0 to sevens, eights, and nines. The Hi-Lo assigns 0 to eights. As sevens and nines have relatively little impact, the two counts are essentially the same." From an examination of Dubner's "Basic Hi-Lo Strategy" 3 × 5 card reproduced in Appendix IV, it seems that the Hi-Lo also assigns 0 to sevens and nines.

Julian Braun first contacted Thorp after the publication of the first edition of *Beat the Dealer* in 1962. He wrote to Thorp and asked him to send a copy of his computer program, which he did. My writing in Chapter 10 that he met Thorp at the American Mathematical Society meeting in 1961 was motivated, again, by a desire for continuity in the exposition.

In a number of references, Julian Braun is referred to as Dr. Julian Braun. Ph.D's were not awarded at the time of his graduate studies in the 1950's at San Diego State College. As noted on the cover of his 1980 book, *How to Play Winning Blackjack*, Braun's terminal degree was the M.S.

A remarkably large number of figures in the history of blackjack, and in Thorp's story, are Jewish. These include Herbert Maisel, Jess Marcum, Harvey Dubner, and Richard Epstein, another 1950's basic strategy scholar, mentioned above. Major characters in the development of Thorp's card counting story were Jewish, including Vivian, Julian Braun, and Manny Kimmel. I accordingly happily include Yiddishkite and Jewish references, most notably in the Braun household of Chapter 2. I have no knowledge of the familiarity of Vivian's parents with Yiddish as described in Chapter 11.

Claude Shannon, the renowned developer of information theory, was a colleague and friend of Thorp at MIT, sharing a well-documented interest in roulette and together developing a wearable computer to win at the game. Thorp enlisted Shannon's aid in deciding which gambler's offer to accept as bankrolling his gambling experiments. In addition, Professor Stewart Ethier informed me that Shannon aided Thorp in publishing his 1961 paper in the *Proceedings of the National Academy of Science* following the American Mathematical Society presentation and recommended a more genteel title for it, "A favorable strategy for twenty-one" instead of "A winning strategy for blackjack." As with most journals published by scholarly societies or academies, an article by a non-member or non-fellow can only be accepted for publication if a member or fellow of the society or academy "communicates" it to the editor. Shannon was the only faculty member in mathematics at MIT who at that time was a member of the National Academy of Sciences. Thorp asked him to so communicate his paper for publication and Shannon agreed to do so. I opted not to clutter the narrative with these side stories. The novel, after all, is a love story between Thorp and Vivian.

Academia and Astronomy

Academics receive notice of appointments in various ways. Most frequently they apply for a position and are directly notified of their success. In many instances, however, a faculty advisor or colleague informs a candidate of a position and facilitates the hiring. This occurred to me personally in my academic career four times. I therefore felt justified in relating Thorp's learning of a position at New Mexico State in Chapter 18 by the chairman of his department at MIT, a choice that enabled progression of the action.

I am happy, as an astronomer, to note that relating Thorp's interest in this field was not invented. Under "interests," Thorp's personal website lists only one, astronomy. When he obtained wealth from his venture into the stock market, his expansive California home included a swimming pool, tennis court, and well-equipped observatory. I should reassure the hoards of astronomers who are reading this book that indeed Venus had passed eastern elongation, as I noted in Chapter 14 during the Nevada trip in the spring of 1961.

The Time Sequence of Events

The time sequence of events was somewhat altered. Table A.3 presents relevant dates, including those true to the history. The first few dates were altered so that Vivian and Thorp could meet in the novel before the date of their marriage. The gambling trip dates were changed so that the trip to Puerto Rico in the novel would follow soon after the trip to Nevada.

As I noted above, at the 1963 Fall Joint Computer Conference in Las Vegas, a panel session was devoted to use of computer simulations to study games of skill and chance. It was there that Dubner introduced the "Hi-Lo" system. I chose not to discuss the conference itself in the novel to avoid cluttering the narrative. Instead, in Chapter 11 I simply had Braun and Thorp refer to Dubner's approach and results. Appendix IV presents an extensive discussion of Dubner's professional life and his approach to the blackjack problem.

Table A.3 A comparison of historical dates in Thorp's life with corresponding or fictional dates related in the novel for exposition purposes

Event	Date in novel	Historical date
Berkeley Jazz Festival	April 1955	didn't occur
Baldwin paper	Early 1955	Fall 1956
Basic strategy trip	January 1956	Xmas 1958
Marriage of Thorps	January 1956	January 1956
Awarding of Ph.D	June 1958	1958
D.C. math meeting	January 1961	January 1961
Nevada trip	Spring 1961	Spring 1961
Puerto Rico trip	Summer 1961	April 1964
Starts at New Mexico State University	Fall 1961	September 1961
1963 Fall Joint Computer Conference	Not discussed	November 12–14, 1963
To Tell the Truth t.v.	February 1964	February 3 1964

Cars, Colleges, and Family Names

Thorp was in fact raised until a young age on the northwest side of Chicago, only two and one-half miles north of the author's home. The Thorp family, Edward, his younger brother James, and his parents moved to Los Angeles in the early years of World War II, before he was ten years old. He acknowledges James in the 1966 second edition of *Beat the Dealer* for the long hours he spent playing, along with Vivian, the role of "the house" in his blackjack experiments.

Specific other elements from the chapters that I created under poetic license follow. I don't know what kind of car Thorp drove, but I doubt it was a red Thunderbird. I have not read that he was a player of darts, for relaxation or otherwise. I invented that to put some color into his fictionalized personality. The "Vivian Name Game" that I introduced in Chapter 5 was manufactured. I hope they realized the richness of this possibility and in fact played it in some form by themselves or with their children. Vivian did have a design business in Newport Beach, but I was not able to determine whether it was fashion, interior, jewelry, furniture, or other type. I chose fashion design in Chapter 5 in the context of her being a creative person in introducing the Vivian Name Game. That seemed the most reasonable for a college undergraduate. Also in Chapter 5, it was the author's father who played the harmonica and liked the tune "Jimmy Crack Corn" and the author's mother whose favorite tune was Artie Shaw's "Begin the Beguine." I doubt that Thorp spent his honeymoon night, as related in Chapter 8, dealing himself blackjack hands.

In Chapter 5 and elsewhere, I said that Thorp was an undergraduate at my *alma mater*, Cornell University. Thorp never attended Cornell University. I know the campus and its area well and wanted to include some of its features. I lived on Glen Place. Again, write about what you know. In particular, when I learned that the academic buildings at both UCLA and MIT have such sterile names, I wanted to juxtapose them for humor next to the Ivy-sounding names of the buildings at Cornell. Thorp, of course, after spending his first year at the University of California, Berkeley, and then transferring to UCLA, obtained all his degrees at UCLA, the B.A. in physics in 1953 and the M.A. in physics in 1955, and the Ph. D. in mathematics in 1958. I fictionalized the location of their home in Brookline during the MIT years. In fact, they lived in Cambridge, to me not as romantic a location.

I have no information on the details of Thorp's dissertation defense and celebratory luncheon as portrayed in Chapter 9, but the elements presented are typical of such events in American academia. Larry Toy, of the luncheon restaurant, is the name of an astronomer who was a graduate student with me at Berkeley. I enjoyed a *dim sum* celebratory lunch in San Francisco's Chinatown courtesy of Dr. Al Cheung of the Berkeley Physics Department after my getting the Ph.D. Al worked with Nobel Prize Winner Charles Townes, who was also on my thesis committee, and with our radio astronomy group and I affectionately placed Thorp's celebration in the same context.

Once he had become known, Thorp did adopt disguises to enable him to continue playing, but he didn't adopt the Chinese disguise described in Chapter 16. That concept was suggested by Thorp's relating the story in the 1966 edition of *Beat the Dealer* of a 1950s systems player who received a Chinese makeover from a Hollywood movie studio make-up artist. Jeffrey Lyle Segal is the name of a make-up man in Chicago with whom I have worked as an actor.

In Chapter 21, I provided the name of their first-born as Fredric, after the famed mathematician and astronomer Karl Friedrich Gauss (1777–1855). In fact, their first-born was a girl. I used a name for the son, which is not the name of one of their children, out of deference to Thorp's personal request for privacy. I don't know the name of any dog or any other animal for whom the Thorps may have provided a home, but, eccentric that I am, I thought I'd name both his fictional son and his fictional dog after Gauss.

Appendix II

Card Counting Systems

As Julian and Thorp foresaw in Chapter 11, various students developed card counting systems following the publication of the simple and complete point count systems by Thorp in the 1966 edition of *Beat the Dealer*. These are based on assigning different point values to the various cards in order to create greater discrimination on the effect of their being dealt from the deck.

Table A.4 provides the most popular and most powerful systems listed in order of complexity. I list them in order of complexity rather than the puzzlingly non-instructive practice of all other gambling writers of listing them in alphabetical order. That makes as much sense as listing the American presidents in alphabetical order. (That would result, I note, in having George Washington, the first president, listed second to last, just before Woodrow Wilson). The numbers in the top row of the table identify the type of card, tens, Jacks, Queens, and Kings being grouped together as *ten-value* cards in blackjack. The numbers in the succeeding rows of the table are the point values assigned to the given card in the top row. In card counting, the total *point count* of a hand is the arithmetic sum of the point values assigned to the cards being held in the hand. If a given developer used pseudonyms, his actual name is provided as a footnote.

As the first widely used system, as well as the currently most widely used system, we provide first the complete point count system of Thorp as published in the 1966 revised edition of *Beat the Dealer*. He acknowledges the contribution of programmers Julian Braun and Harvey Dubner in developing this system, often referred to as "high-low" or "hi-lo." Dubner's significant role was discussed in Appendix I and is more fully discussed in Appendix IV.

© Springer International Publishing AG 2017
L.M. Golden, *Never Split Tens!*, DOI 10.1007/978-3-319-63486-9

Table A.4 The most popular card counting systems are presenting in order of complexity

Name	Acronym	2s	3s	4s	5s	6s	7s	8s	9s	10s	Aces	Developer
Complete Point Count	High-Low	1	1	1	1	1	0	0	0	−1	−1	Edward O. Thorp [1,2]
Revere Plus-Minus	Revere +/−	1	1	1	1	1	0	0	−1	−1	0	Lawrence Revere[1]
Knock-Out	K-O	1	1	1	1	1	1	0	0	−1	−1	Olaf Vancura/ Ken Fuchs
Silver Fox		1	1	1	1	1	1	0	−1	−1	−1	Ralph Stricker
Canfield Expert		0	1	1	1	1	1	0	−1	−1	0	Richard Canfield[3]
Hi-Optimum I	Hi-Opt I	0	1	1	1	1	0	0	0	−1	0	Lance Humble/Carl Cooper
Uston Advanced Plus/Minus	Uston APM	0	1	1	1	1	1	0	0	−1	−1	Ken Uston[1]
Keep It Simple 2	KISS 2	0/1	1	1	1	1	0	0	0	−1	0	Fred Renzey
Keep It Simple 3	KISS 3	0/1	1	1	1	1	1	0	0	−1	−1	Fred Renzey
Red Seven		1	1	1	1	1	0/1	0	0	−1	−1	Arnold Snyder[1]
Advanced Omega II		1	1	2	2	2	1	0	−1	−2	0	Bryce Carlson
Canfield Master		1	1	2	2	2	1	0	−1	−2	0	Richard Canfield[3]
Zen Count		1	1	2	2	2	1	0	0	−2	−1	Arnold Snyder[1]
High Optimum II	Hi-Opt II	1	1	2	2	1	1	0	0	−2	0	Humble/. Cooper
Unbalanced Zen II	UPZ II	1	2	2	2	2	1	0	0	−2	−1	George C.
Mentor		1	2	2	2	2	1	0	−1	−2	−1	Fred Renzey
Revere Point Count	RPC	1	2	2	2	2	1	0	−1	−2	−2	Lawrence Revere[1,3]
Uston Advanced Point Count	Uston APC	1	2	2	3	2	2	1	−1	−3	0	Ken Uston[1]
Uston Strongest & Simplest	Uston SS	2	2	2	3	2	1	0	−1	−2	−2	Ken Uston[1]
Revere 14 Count		2	2	3	4	2	1	0	−2	−3	0	Ken Uston[1]
Revere Advanced Point Count	RAPC	2	3	3	4	3	2	0	−1	−3	−4	Lawrence Revere[1,3]
Wong Halves		0.5	1	1	1.5	1	0.5	0	−0.5	−1	−1	Stanford Wong[1,3]

[1]Member of Blackjack Hall of Fame
[2]Thorp acknowledges Harvey Dubner and Julian Braun as significant contributors to the complete point system
[3]Pseudonyms (given name): Richard Canfield (John Hinton), Lawrence Revere (Griffith K. Owens), Stanford Wong (John Ferguson). Owens also traveled and gambled under the aliases of Leonard Parsons and Specs Parsons.

In what I consider to be an unwarranted and unnecessary attempt to impart mathematical academic *gravitas* to a trivial point, some employ the term "level" to refer to the absolute value of the largest point value utilized in a system. In this way, for example, in a so-called level 1 system the point values assigned to the cards are either -1, 0, or $+1$. I find that term unappealing. It suggests a superiority to the systems with the highest possible point values, which is not necessarily the case, other factors and criteria influencing such a judgment. They are certainly more difficult to learn and use at the blackjack tables. I prefer the simple descriptive term *complexity n* or *complexity value n* system. This implies the increased difficulty of using systems with maximum absolute point value greater than 1.

Billions of possible systems could thereby be developed. With choices of -1, 0, and $+1$ for the simplest systems, for example, one has $3 \times 3 \times 3 \times 3 \times 3 \times 3 \times 3 \times 3 \times 3 \times 3 = 3^{10}$ possible systems, about 59,000. Similarly, about 9.8 million, 280 million, and 3.5 billion mathematically possible systems exist for the systems with maximum point values of 2, 3, and 4, respectively. The systems of different complexities displayed in the table are distinguished by the color of their cells. I distinguish the Wong Halves system, in which non-integer point values are utilized, from the rest.

Because an effective card counting system depends on the low value cards being assigned positive point values and the ten-value cards and Aces generally being assigned negative point values, many of these possibilities are immediately eliminated as not viable. Among, for example, the ten systems in the least complex, simplest, class, only the twos, sevens, nines, and Aces differ in point values. Still, that allows $3^4 = 81$ possible systems.

Certain trends in the table are obvious and provide insight into the difference between systems. Far more complexity 1 systems have been developed than complexity 2 systems, and more complexity 2 systems have been developed than the more complex systems. Among the complexity 1 systems, all agree for the point values ascribed to six cards, the threes, fours, fives, sixes, eights, and ten-value cards. Among the complexity 2 systems, agreement holds only for five cards, the fours, fives, sevens, eights, and ten-value cards. The more complex systems show less agreement between systems.

These trends can be explained on the basis of simplicity versus precision and gambling philosophy, or goal, of the developer. The easiest systems to use are the simplest, which are therefore likely to gain popularity. This leads developers to focus on complexity 1 systems. The more complex systems more precisely discriminate the effect of the given card being dealt from the deck. Because a careful player would expect to play more profitably if these could be mastered, they are targeted to those who have mastered the complexity 1 systems. The differences between systems of a given complexity result from the desire of the developer to provide the player with a greater advantage based on either betting, strategy, or to a lesser extent insurance strategies. In actual play in casinos, the errors made by players can easily mask any difference in these advantages among systems of a given complexity. As a result, simplicity of the system most frequently determines what system a given player will adopt. Thorp's complete point count system remains, as noted, the most widely used. For the record, the author uses only the complete point count system of Thorp, Dubner, and Braun.

OOPAH!

Casino Reaction, Dubner's Personal Blackjack
Winnings, and His Personal Reflections

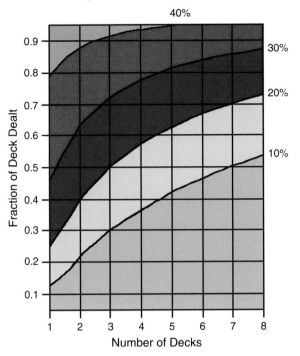

The "Golden Diagram" for a high-low index of +10. The contours show the probability of attaining a value of the high-low index of +10 as a function of the fraction of the deck dealt and the number of decks in the deck pack. Casinos worldwide adopted the use of multiple decks instead of only a single 52-card deck after the development of card counting and its popularization by Edward Thorp. Mathematicians then began to develop techniques, such as that based on Golden Diagrams, to ensure that the game of 21 would still be profitable to card counters. The formula for the high-low index is discussed in Chapter 11. The theory behind the Golden Diagram was developed in a series of articles in *Bluff Europe* magazine and published in the peer-reviewed *The Mathematical Scientist* in 2011.

References

C., George, *The Unbalanced Zen II* in Schlesinger, Don (2005) *Blackjack Attack: Playing the Pros' Way*, RGE Publishing, Ltd., North Las Vegas.

Carlson, Bryce (2010) *Blackjack for Blood*, Gamestar Pr Inc.

Humble, Lance and Cooper, Carl (1987) *The Worlds Greatest Blackjack Book*, Doubleday, New York.

http://en.wikisage.org/wiki/Blackjack_Card_Counting_Systems.

Renzey, Fred (1997) *Blackjack Bluebook*, Blackjack Mentor Press.

Renzey, Fred (2006) *Blackjack Bluebook II*, Blackjack Mentor Press.

Revere, Lawrence (1971) *Playing Blackjack as a Business*, Carol Publishing Group-Lyle Stuart, New York.

Snyder, Arnold (2005) *Blackbelt in Blackjack*, Cardoza, Las Vegas.

Stricker, Ralph, *Silver Fox Blackjack System*, Gambler's Book Club, Las Vegas.

Thorp, Edward O. (1966) *Beat the Dealer: A Winning Strategy for the Game of Twenty-One*, Vintage, New York.

Uston, Ken (1998) *Million Dollar Blackjack*, Carol Publishing Group, New York.

Vancura, Olaf and Ken Fuchs (1998) *Knock-Out Blackjack*, Huntington Press, Las Vegas.

Wong, Stanford (1994) *Professional Blackjack*, Pi Yee Pr Press, Las Vegas.

Zadehkoochak, Mohsen (1992) *The Book of British Blackjack*, Medimage Books, Barnet, North London, United Kingdom.

Appendix III

Camouflage: An Acting Primer for Card Counters

Lesson One: Disguises and Costumes

Hiding your being a card counter or a member of a blackjack team of card counters is just as important as the strategy techniques you employ in table games. What does it matter if you're the most skilled card counter if your style gives you away and you are banned from the tables? You must also prevent detection by the casino personnel, the dealer, the pit boss, and most importantly, the guys behind the one-way mirrors in the ceiling and watching the closed-circuit TV screens in the security rooms. You must be aware of this need to avoid detection between different shifts at the same casino, between different casinos in the same casino town, and between the groups of casinos in different casino towns.

I categorize the various camouflage techniques in blackjack using the acronym *ABS*. That stands for "acting, betting, and strategy." The "acting" camouflage involves pretending you're not a counter but, for examples, a skier, partier, showgoer, or mortgage-backed securities broker joke-teller, although none of these exists. The "betting" and "strategy" aspects involve making bets and splitting, doubling down, drawing and standing, and insurance decisions contrary to those that an effective system would dictate.

I could have chosen a different acronym created by permuting the letters, such as BAS, but that's the suffix for programs written in the BASIC programming language; BSA, but that's reserved for the Boy Scouts of America, who DO NOT GAMBLE; or SBA, but that's the American "Small Business

Association," and believe me counting cards is NOT a small business. The other two were also unavailable for my use, ASB, the Aphrodisiac Society of Bulgaria, and SAB, the Strategic Aviation Bottleneck, the heralded protective umbrella over Egypt created by the vaunted Egyptian Air Force.

In the novel, numerous betting and strategy camouflage techniques were introduced. Here I want to discuss the acting element.

Although acting is the least appreciated element of camouflage strategies, it's crucial, and I list it first in the acronym. I learned many acting techniques from my good friend and badminton partner Clint Eastwood, who refers to me as "the guy who always cracks me up on the set with his Kama Sutra-origami designs. What a guy!" Although this book concerns the game of blackjack, acting techniques are also crucial for members of roulette teams and poker teams, whose members can share the identity of their hole cards with each other by verbal codes.

The acting, then, can camouflage your being a blackjack card counter or being a member of a blackjack, roulette, or poker team. In the following, I'll generally refer to an individual card counting at blackjack, but the techniques can be used by team members as well.

Because many of the elements of acting are irrelevant to the action at the tables, this is obviously not going to be an acting class. I'm going to restrict myself to lessons in those areas of acting that I think are relevant, that is, disguises and costumes, speech accents and dialects, communication of emotions, and relationships. The first two are easily employed; the latter two require some acting skill.

The casinos share photos of suspected counters. If you win consistently, especially in large amounts, your description and photo will be communicated around to the casino personnel and to other casinos. Once you're detected, usually by your table manner, betting pattern, and strategy choices, the casino has various ways of nullifying your game. The dealer may be told to shuffle up the deck on you when you increase your bet. The casino might put a shill next to you who won't allow you to see his hole cards, replace the dealer with a cheating dealer, a "mechanic," or simply bar you from playing and plying your trade. If you walk in and sit down, they might get a security guard to escort you out. "We don't want your action here." They'll threaten to have you arrested if you return. I know; I've received those threats and have enjoyed the company of escorts. These aren't the kind who meet you for a dinner and a show. Boors!!

As a result, once detected, you have to either travel to another gambling mecca or, if you want to continue to play in the given casino town, hide who you are. That's where disguises and costumes come in. They hide your identity so that you can continue counting in the same casino at different shifts, at different casinos in a given casino town, or at casinos operated by the same company in different casino towns in, say, Nevada.

Changing your physical appearance by wearing costumes and using makeup, wigs, and beards is one element, of course, of the ancient trade of acting. In gambling, we don't really talk about costumes in of themselves. The emphasis historically has been on disguises, and within that context we can include elements of traditional theatrical make-up and costumes. This adoption of disguises has a colorful anecdotal history in card counting, ranging from simply growing a beard to full-blown Hollywood makeovers.

Thorp in the 1966 second edition of *Beat the Dealer*, for example, recounts his inability to "get a reasonable game" after the intense publicity and wide distribution of his photo following the success of its first edition in 1962. "As a last resort," he adopted a disguise for a four-day visit to Las Vegas, Reno, and Lake Tahoe. He grew a ragged beard, abandoned his eyeglasses for contact lenses, wore wraparound sunglasses, and put on nondescript clothing. He was able to win for two days in Las Vegas, but then bearded players, having won consistently, began to receive hostile treatment.

He moved on to Lake Tahoe and then Reno. The word, however, had spread among the casino towns to look out for a bearded man. He continued to win, but in both Lake Tahoe and Reno the casinos replaced their dealers with mechanics, in Lake Tahoe upon observing the bearded player and in Reno simply because of his winning. None, however, recognized him as Thorp of *Beat the Dealer*.

His final evening trying to hide his identity, in Reno, he abandoned the bearded disguise for a second one. He shaved, removed the sunglasses, wore a suit, and combed his hair in a different style. He proceeded to win again, no longer attracting attention as the bearded man.

If more elaborate disguises appeal to you, visit your local theatrical make-up store, magician's store, or the make-up section of your local drugstore. I've found it best to shop preceding and, for bargains, immediately after, Halloween. You're not going to go to the casino as Count Dracula, Count Alucard, or Bozo the Clown, but looking at the various gadgets and accoutrements will give you some ideas.

Rather than adopting a specific disguise, you can also simply go to the casino dressed appropriately to the region, season, and time of day. Don't, that is, dress in the casual outfits of many card counters, and certainly don't wear a tweed coat and bow tie even if your life's wish was to become a math-whiz professor. Camouflage your counting by the clothes you wear. These are effectively the actor's "costume."

In Lake Tahoe during the winter, wear your ski jacket to the table and put rouge on your cheeks. Then talk about your skiing. While rubbing your hands, say, "Boy, I just can't get warmed up."

In Las Vegas, if it's the morning or afternoon, wear your golf outfit and talk about your score. Bore the dealer with conversation about your new wood. If you're playing after dinner, wear a suit and say you just came from that hot new restaurant. Alternatively, wear a coat but no tie, and ask the dealer if ties are required to dine at their high-end French restaurant or to see Wayne Newton or Shecky Greene. Call the pit boss over to see if he knows. You might also wear a coat and tie and discuss the show you're going to see. After all, you're at the table simply to pass the time until dinner or the show begins, right? If you're there looking for the wild Vegas parties, wear a couple pounds of gold chains, open-chest shirt, some large pinkie rings, and a fake Rolex watch or two.

In other words, fit in. Appear to be the type of person that the casino is accustomed to see at the tables.

Wearing such outfits and discussing such subjects will help persuade them you're not a counter, but only at the tables to pass the time until you can warm up, get on the links, dine, or get to your show or babe-infested penthouse.

That's one element of disguise, to hide your identity so that you can continue card counting in the same casino at different shifts, at different casinos at a given casino town, and at casinos operated by the same company in different casino towns. It's also important for the members of blackjack, roulette, and poker teams to dress differently. The casinos obviously also share photos of suspected team members.

If you all go dressed in jeans and gym shoes, and you're all wearing cell phone holsters on your belt and MIT calculators on your neck chain, you might as well be wearing team uniforms. "Here we are, Mr. Pit Boss. The MIT Card Counting Team. I'm number 12; see the number on my jersey?"

The members of the team should be dressed in different "costumes." In summer, one could be in a gray suit, another in a blue tennis outfit, and another in a black and white casual shirt. In winter, the outfits could include a burgundy corduroy suit, a yellow ski parka, and a brown and red woolen shirt with a red woolen scarf.

I noted that you must be aware of the need for camouflage not only between different shifts at the same casino but also between different casinos in the same casino town and between the groups of casinos in different casino towns. Outfits greatly aid in this.

Just as actors have different outfits in different scenes in a movie or different acts in a play, bring along different outfits on your gambling trip. If your team has done well at one casino, the word will get out to the other casinos to be on the lookout for guys in a burgundy corduroy suit, yellow parka, and brown and red woolen shirt and red scarf. Vary the outfits you wear between the various casinos. If you have done well and plan to hit another casino town, perhaps consider a second complete set of outfits.

Lesson Two: Characterization – Speech and Emotions

"Yonder Lies the Castle of My Fodder"

A life in the theatre is so glorious, so splendid – I do think it's the most wondrous of callings, isn't it? Unless, of course, you use those same acting techniques to rip off a few gazillion in the casinos by, for example, using a British accent.

Becoming someone other than yourself is a challenge to all actors, and many methods and techniques have been developed over literally hundreds of years to accomplish that end. Actors spend their entire careers refining this skill, but, again, this is not an acting class. I'm not going to discuss Stanislavski, the "method," the techniques of the actors' studio, "sense memory," motivation, sexual addiction, and other subjects.

Just seeing if you're paying attention!

I'm not going to discuss all these things because, after all, when you're sitting at the tables you often speak very few words. The only subjects I will discuss are the accents and dialects and emotion aspects of characterization. These are valuable elements of the "acting" part of my *ABS* camouflage technique of "acting, betting, and strategy," valuable because you can use them at the tables.

Accents and Dialects

That guy in the burgundy corduroy suit could speak with a French accent, the gal in the yellow parka could speak broken Italian, and the man in the brown and red woolen shirt and red scarf might want to order matzo ball soup at the table.

Many stage and film actors have learned to speak with some of the most commonly used accents such as upper-class British, Cockney, French, Irish, Italian, Southern U. S., Japanese, Jewish/Yiddish, and Russian. Many have also learned the speech patterns common to American Negro/jive, gangster/Chicagoese, and "flaming gay." The early work of the late Tony Curtis became legendary for him supposedly bringing his Bronx accent to a Medieval character, although the story is incorrect; Curtis never in fact stated the line that's part of the title of this essay.

Some good books describe the vowel and consonant substitutions, and some, including Jerry Blunt's *Stage Dialects,* are accompanied by recordings of native speakers. The Internet is also a good source for such material.

Accents can magically transform you into the character of that speaker. Saying, for example, "Hey, youse gonna giv' me som' good cards, or ain't ya'?" in gangster/Chicagoese nearly forces you to adopt the poorly educated, rough street persona of the punks growing up in the neighborhoods of Chicago. Asking, "Oh, really, would you be so kind as to deliver an Ace for me, would you?" in an upper-class British accent makes you adopt a gentile and cosmopolitan nature. How about, "I like-a you. You come-a to my house-a and my wife-a Francesca she-a make-a for you some of her famous-a pasta, hey?" That homey warmth is not spoken with a Swedish accent, paison.

You don't have to become actors and spend an inordinate amount of time studying dialects. Just a few substitutions of pronunciations and a few key words and phrases will identify your adopted nationality to the dealer and pit boss. To show you that this is easily done, I'm going to provide some hints on speaking with French, Yiddish/Jewish, and Italian accents.

For French, substitute the vowel sound of "eat" for that of "it." The letter "h" is not sounded either in the initial or medial positions. In this way, "hit" and "perhaps" become "'it" and "per-'aps." The former is particularly suited to black-jack, when asking for a card, "Please, 'it me." The sound of "z" is substituted for the "dz" sound in words such as "justice" and "page," and it's easy to extend that "z" sound to other words. Using some common French words also help. Use "si vous plait," "oui," and "monsieur" in your speech.

You get a blackjack. Inform them you're a French tourist by exclaiming, "Zis blackzhack. Eet iz zo beautiful. I am zo 'ot."

Jewish/Yiddish is another easily adopted accent, and it's fun to use. The short "i" as in "it" becomes "e" as in "eat," the word "a" is elongated to "ahhh," and the final "g" as in "counting" becomes a "k." Also, at the beginning of words change the sound of "w" into a "v" and that of "th" into a "d." For example, "It's a good wife, cooking this chicken soup" becomes "Eet's ah good vife, coookink dis chicken soup." When you come to a table and someone is drinking a crème de cocoa you can provide some culinary information. "You vant ah leetle advice? Never you should be drinkink ah milkshake vit matzo ball soup."

The Yiddish language is full of words beginning with "sch" or "sh" such as "schmooze," "schlemiel," and "schmuck." Other words are misused, for example, "mein" instead of "my." You can begin your conversation with the dealer simply by asking, "So how's the family? You shouldn't know from my problems and dey start vit mein brother-in-law Marvin, dat schmuck." When the waitress comes by, throw in, "You know vat? Dat's ah cute schicksie."

The most distinctive sound in Italian is the use of the letter "a," pronounced as in "up." Every noun and the overwhelming majority of other Italian words end in a vowel, as opposed to a consonant. As a result, Italians speaking

English find it unnatural to end a word with a consonant, especially if that word ends in a vowel in Italian, and append that "a" sound. When the native Italian speaker is speaking English, this sound appears as the final sound in a word ending in a consonant, as in "you wanna buy-a dis-a cap-a?," between two consonants in a multisyllabic word, as in "mid-a-night," and between two words when both the first ends and the second begins with a consonant, as in "what-a she want-a?"

English contractions such as "can't" and "don't" provide trouble for Italian speakers and are simplified by simply using the words "no" or "not." "He no can do it." "She not gonna take a hit on 18." Your Italian heritage becomes established if you throw in some homey expressions such as "You come over for-a poker-a game at-a my-a house-a some-a-time, okay?" or "Your-a brother, he got-a meet-a my-a cousin-a, Tina."

To be complete, I should note that the imposition of disguises and costumes also serve the purpose of aiding characterization. Just as with accents and dialects, adoption of disguises and costumes help transform you into a character.

Remember to use different accents and dialects between different shifts at the same casino, between different casinos in the same casino town, and between the groups of casinos in different casino towns.

Keep your goal in mind always. You want to camouflage your being card counters or members of teams. The dealer and pit boss will think that an Italian shoe repairman and a debonair French man are on vacation, hardly card counters, and strangers, hardly members of a blackjack, roulette, or poker team.

Communicating Emotional States and the Subtext

Acting books and instructors generally do not consider the communication of emotions as an element of characterization. Rather, a character can experience different emotional states. Here, however, we can utilize an emotional state to camouflage our identity at the table. The dealer or pit boss will remember the guy who was angry or happy, and for him that is your character.

By emotion I mean your current psychological state. Joy, anger, excitement, passion, fear, sorrow, sexiness, eagerness, regret, compassion, and other emotions have been experienced by all of us. Practice being in such states. A good exercise is to take a phrase and deliver it with the different emotions. Carroll Anderson, my high school public speaking teacher, used the phrase "Bob, I didn't know you could do it." (You'll remember him from Chapter 14 as the teacher who assigned an exercise to describe something beautiful without actually using the word.) When it was my turn, by the way, my interpretation got a laugh from everyone. I don't know what I really meant, but it was some sort of "dirty" romantic gymnastics. You could use others with even less content. "That's the way it goes" or "tomorrow will be a good day" would work. Keith Johnstone in *Impro* presents an exercise in which the actors communicate their state of mind by uttering, "It's Tuesday." You can even try, "It's your turn to take out the garbage."

The trick I've used to great success as an actor, if I may be so bold, is to say a phrase silently to yourself that clearly communicates your feeling, and then blurt out the phrase. If you want to express, for example, fear by stating "Bob, I didn't know you could do it," then say silently to yourself, "The giant spiders are knocking down the door and they're going to eat us all," and then blurt out, "Bob, I didn't know you could do it." After just a little practice, it'll come out as "fear."

For another example, say to yourself, "I won the lottery! One-hundred million dollars! I'm rich," and then state, "It's your turn to take out the garbage."

In fact, the words are simply the carrier wave for the emotion. You can speak gibberish and still express the same emotion. Try saying silently to yourself, "The giant spiders are knocking down the door and they're going to eat us all," and then blurt out, "Ish ga flunken ka bibble." With some practice it'll come out as fear, unless of course you have a thing for spiders.

Also be aware that changing the volume of your speech aids in expressing the emotion. Joy and anger, for example, are often spoken loudly. Sexiness, regret, and compassion are spoken softly. What a rotten choice it would be if you yelled, "I'm so sorry your wife and daughter got scurvy in that casino in Macao. I'm here for you." It would be just as silly to whisper, "We won! We won the World Cup and now I'm going to party till dawn!"

Acting books and instructors refer to the *subtext* as the actual message being communicated. A major element of that subtext is the emotion you are experiencing.

Now, to put this in practice at the tables, adopt a character's emotion. Every time you address another player or the dealer or croupier, use that emotion, be it fear or excitement. You should adopt different emotional states between different casinos in the same casino town, and between the groups of casinos in different casino towns.

Don't get carried away with this. Acting books and instructors who know what they are talking about will tell you that you cannot act a state of mind or emotion, but can only perform an action. You are directed to act out, that is, manifest your emotional states. In film and on the stage, this can be done by physical action as well as modulating your voice. At the casino table, your physical action is constrained by your sitting. You are limited in your actions.

You really don't want to attract attention to yourself by jumping up and screaming in joy, or pounding the table in anger. Our goal, after all, is to not attract attention and scrutiny. Those same acting instructors will tell you not to describe your emotional state. Uttering statements such as "I'm so angry," "I'm so happy," and so on, in film and on the stage, does nothing to progress the action, and such phrases are avoided by good writers. At the casino table the concern is that such statements may seem phony.

Lesson Three: Relationships

"The Rain in Spain Falls Mainly on the Plain"

Well, actor-counters, by now you're accomplished theatrical performers, but one lesson remains. To prevent being barred from plying our trade, we want to hide from the house our being card counters or members of blackjack, roulette, or poker teams. To do this, I use my ABS camouflage approach of "acting, betting, and strategy." In addition to the disguises and costumes, accents and dialects, and the communication of emotions aspects of acting, you can camouflage your play by clearly defining your relationship, in our present context, how you relate to the others at the tables. Actors with traditional training often refer to their *attitude*, which is basically one important aspect of relationship. After all, one's attitude occurs in the context of relating to another person or, less frequently, object or concept. Accordingly, I present one more aspect of acting for gamblers, the all-important *relationship*. Like, "So, do you feel lucky, punk? Well, do ya'?" or "vere's da #$%@ matzo ball soup!"

   A relationship determines how two people interact. At the tables, you can have multiple relationships, and they can aid your camouflage. The relationship can be with your friend who is gambling with you or standing behind you, the dealer, the pit boss, another player, or the cocktail waitress.

Defining a relationship is less difficult for actors than developing a character. After all, the playwright tells you if so and so is your solicitous mother, distrustful business partner, or loyal butler. At the table, too, the relationships are often right there for you. You're envious of the high roller. You lust after the cocktail waitress. Your girlfriend provides encouragement. You worry about the pit boss checking you out.

The single most important factor defining the relationship between two people is their *relative status*. Because the concept of relationships extends far beyond blood relationships, we want to use this more general term. The word "relationship" comes from the same root as the word "relative" anyway, I guess. This term is also useful because many of the elements that define the status between two people are relative measures, and it would be confusing to discuss "relative relationships."

Factors creating status include age, physical appearance, education, wealth, power, happiness, honor, ambition, knowledge, cunning, experience, wardrobe, grooming, emotional stability, material possessions such as automobiles and jewelry, and voice and speech quality, that is, diction, accents, and vocabulary. So we see that disguises and costumes, accents and dialects, and communication of emotions, the elements of acting we've discussed before, all help establish your relative status.

Instead of using the term "relative" status, I sometimes prefer the term *differential status*. Then we can use the concept of large as opposed to small differentials to provide some concept of quantification, which has advantages in providing guides to behavior. Two characters of small differential status, for example, two benchwarmers on a youth football team, will interact differently from two characters of large differential status, for example, a baron and his serf. Again, if a small differential status exists between person A and person B, and a large differential status exists between person A and person C, then we can say that person B has higher status than person C. You can also compare relative status based on different attributes. Whenever a script has the phrase, "We're really the same, you know, the two of us," you are viewing small differentials in status based on different attributes. Try that sometime with a dealer.

We can now state that an individual's behavior depends on status relative to that of the person with whom he or she is interacting. In camouflaging your play by acting, we want to be aware of this factor and utilize it.

As I noted, many acting books and instructors refer to the attitude that one character has towards another. A large part of that attitude is defined by their differential status. If you are a waiter, then your attitude toward a diner will be of deference. If you are the diner, your attitude may be, if you're not particularly gracious, tolerance. No need exists here to dwell on this concept other than to note it defines differential status.

To show that your differential status with others at the gambling table defines you, here are two examples. Let's say a member of your blackjack team is pretending to be your wife. You can act high status toward her, advising her on her moves. "You never split 10's, honey." You can act low status toward her, her berating you constantly. "You're betting too much. I told you we should just play keno."

You can do the same with the dealer. You could act low status: "Gosh, you must be real smart to be hired to work at the Mirage. There must be hundreds of people you beat out for this job." You could act high status. "You know, I have a 40-acre estate in the Bahamas. I employ twenty people, and they all make double your salary. But I still think you're a decent person. If you're ever interested in leaving this boring job, set up an interview with one of the HR people at my investment firm."

Here are some specific phrases to give you an idea of how you can establish low and high status. To establish your low status, you might reluctantly approach a table and ask, "Is it all right for me to sit here? Is this seat reserved for a high roller?" That implies you're not and that certain seats are reserved for special people, and you're not one of them. You could say, "If I sit here do I have to stay a certain amount of time? I'm just waiting for the Mrs." That shows again you're concerned about violating "rules" and, more, your wife is the boss.

When you're playing, be apologetic. Ask for a card by saying "hit me" rather than scratching the table with your cards. "Oh, I'm sorry, I forgot to ask for a card in the right way." If you get blackjack and throw it down on the table in the middle of another player's play, be effusive. "Oh, I'm sorry. I hope I didn't break your concentration. I'll wait my turn next time."

Low status doesn't necessarily mean wimpy. If a guy sitting next to you is well-dressed, you can admire his outfit. The implication is that you can't afford the same. If the cocktail waitress is a fox, you can say, "I've dated some nice girls, but nobody who looks like that."

To establish your high status, you can similarly use appropriate phrases when you approach the table and during play. "I'm going to sit at this table, but I don't want any smokers here." "Is this the seat reserved for studs and important people? Then I'm sitting here." During play, you can put down the casino, the pit boss, the dealer, or the players. "If my carpet installers brought in this kind of carpeting I'd fire them on the spot." "Don't they ever replace these dirty felts?" "The pit boss looks like he could use a real job." "That player hasn't bathed in a week. Don't you have a V.I.P. room?"

Status differentials in the theater can be large or small. In the casino, because you have to establish your rank quickly, go for the large status differentials. This may irritate the dealer or croupier, but that's good. You want him to think of you as anything but a card counter or team member.

If you're a member of a blackjack, roulette, or poker team, large differentials between members will hide your working as a team. The people in teams are usually similar in age, education, and personality and that's what the casinos expect.

As hinted at above, you can also apply status to an object. One example I provided concerned the carpeting and felt. At the tables you can also relate to the chips, dealer outfits, ceiling lighting, casino furnishings and color scheme, and so on.

Defining your status clearly will also prevent your being identified at other shifts in the given casino. If your team member-wife is constantly criticizing you, the description that a pit boss who thinks you may be a blackjack team will provide will include a comment to that effect. When you go to another casino, switch status. Now the "husband" would be the high status actor, criticizing the "wife." Similarly, if you're an overly apologetic counter at one casino, complaining excessively at another ("there's too much ice in this drink") will prevent them from recognizing you.

Remember, you should vary your relationships with other people at the table and with your team members not only between different shifts at the same casino but also between different casinos in the same casino town and between groups of casinos in different casino towns.

Lastly, as with using accents, simply establishing your status can prevent your being discovered as a counter. Many counters sit silently, observing the cards as they fall. Because talking is a major means of establishing your status, you'll be talking and counting at the same time. That helps hide your being a counter.

A Final Word of Encouragement Before the Curtain Goes Up

You should practice these acting techniques with as much dedication as you practice your card counting or sharing of signals with your team members. Wearing a distinctive disguise or costume, equipped with the skills to deliver your words in an accent or dialect, with a given characterization and to a person with a defined relationship, you can camouflage who you really are. You're a counter or team member – one who may decide to embark on a career as an actor.

On the other hand, I hope that, in case you're a gambler, these lessons have not stimulated you to think of abandoning gambling for acting. The money you can make at the tables, with the aid of the camouflage of acting, is far more than the money that 99.9% of all aspiring actors will ever make.

Appendix IV

Harvey Alan Dubner

The Developer of the "Hi-Lo" System

Because of production schedule constraints, the material presented in Appendix IV should be considered a preliminary result of Professor Golden's scholarly research into the contribution of Harvey Dubner to blackjack theory. We regret any errors of commission or omission. The final version of the paper is presented at http://en.wikisage.org/wiki/Harvey_Dubner.

Harvey Alan Dubner was born on July 14, 1928, the son of Reuben and Frances Dubner, New York Jews of Eastern European descent. He had a younger brother, Neil Peter Dubner. He and his wife Harriet (Weiss) had four children, in order of age, Robert Joseph, Emily Rachel, Terry Ann, and Douglas David. Although his entire family was Jewish, his son's recollections below indicate they did celebrate the exchange of gift ritual of Christmas during Hanukkah.

He attended Brooklyn Polytechnic University in New York, from which he earned a Bachelor of Science in electrical engineering in 1949 summa cum laude and a Master's degree of electrical engineering in 1951. He played flippantly with his middle name. Although his birth certificate states his middle name as "Alan," his Master's degree diploma from Brooklyn Polytech in 1951 provides it as "Allen." His son Robert notes that "it is clear that he never knew how his middle name was spelled. He would use his middle initial when forced to. But asked about his middle name, he would shrug and spell out

Allen. But it was always clear to me that he was just coming up with something to make the questioner go away; he neither knew nor cared how it was spelled."

Although educated and employed as an electrical engineer, he was a proficient mathematician and as early as the 1950s a skilled computer programmer. He found great pleasure and joy in attacking mathematical problems working alone. He would first try to solve the problem analytically, using only equations. If the problem was not amenable to such solution, he would write a program to simulate the effect.

His career brought him in 1960 to the Curtiss-Wright Corporation in Paterson, New Jersey. There he used an IBM 1620 computer to run blackjack simulations, leading to the Hi-Lo system. In so doing, he solved the problem of an optimum bet, independently recreating the 1956 work of J. L. Kelly, now known as the Kelly criterion.

He considered the twos through sixes as "lows" and the ten-value cards and Aces as "highs." His analysis then showed that an index calculated by subtracting the number of low-value cards left in the deck from the number of high-value cards left in the deck and dividing by the number of cards remaining to be played provided an excellent sensitivity to the quality of the remaining cards.

Dubner, then 35 years old, presented his blackjack results during a panel discussion on games of chance and skill, entitled "Using Computers in Games of Chance and Skill," moderated by Thorp at the 1963 Fall Joint Computer Conference held in Las Vegas. He had learned of the panel discussion, most likely in the trade magazine *Electrical Engineering,* and had contacted the conference and asked to be included as one of the presenters. He had his results for his betting and playing strategies printed on 3 × 5 index cards. The clamor by those in the overflow, standing-room only crowd to obtain a copy of the "Basic HI-LO Strategy" card caused, in the words of his wife, Harriet, a "riot." Gambling writer Jerry Patterson writes that Dubner was "mobbed." An image of that card is reproduced here.

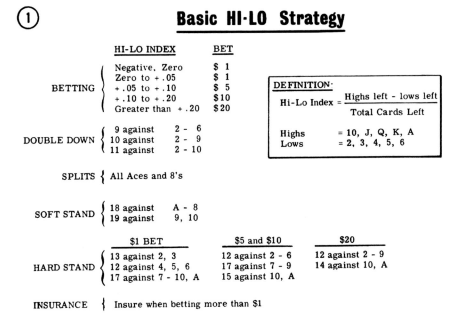

Basic HI-LO Strategy

HI-LO INDEX		BET
	Negative, Zero	$ 1
	Zero to + .05	$ 1
BETTING	+ .05 to + .10	$ 5
	+ .10 to + .20	$10
	Greater than + .20	$20

DEFINITION·

$$\text{Hi-Lo Index} = \frac{\text{Highs left - lows left}}{\text{Total Cards Left}}$$

Highs = 10, J, Q, K, A
Lows = 2, 3, 4, 5, 6

DOUBLE DOWN	9 against	2 - 6
	10 against	2 - 9
	11 against	2 - 10

SPLITS { All Aces and 8's

SOFT STAND	18 against	A - 8
	19 against	9, 10

HARD STAND	$1 BET	$5 and $10	$20
	13 against 2, 3	12 against 2 - 6	12 against 2 - 9
	12 against 4, 5, 6	17 against 7 - 9	14 against 10, A
	17 against 7 - 10, A	15 against 10, A	

INSURANCE { Insure when betting more than $1

A reproduction of the 3 × 5 card distributed by Harvey Dubner at the 1963 Joint Computer Conference in Las Vega. It caused, in his wife's word, a "riot." (Courtesy of Dubner family)

Most, if not all, blackjack historians have assumed that Dubner read the 1962 edition of Thorp's *Beat the Dealer* and was thereby stimulated to examine the game himself, leading to his development of the system. His son, Robert Dubner, in our early correspondences made the same assumption. It was natural to do so. Thorp distinguished tens and non-tens in his ten-count strategy published in 1962, and Dubner distinguishing "highs" vs. "lows""in the "Hi-Lo" system seems a natural extension rather than a complete re-invention. Yet, further discussion with the Dubner family – Harvey, his wife Harriet, and his eldest son Robert – revealed that Dubner did not know of Thorp until the 1963 conference. That he worked independently of anyone else on the project is well known, confirmed by Thorp himself.

An absolute validation of this thesis would be available if the Dubner family had retained some of the paper tape printout of Dubner's computer runs. In the early days of computers, output was provided on paper tape and punched cards. Later computers provided output on magnetic tape and within the computer memory. Those paper tape printouts would have either had the date of the run encoded, or perhaps Dubner identified the date of the run with a marking pen, stick-on label, or a note taped to the paper tape. A date of any run pre-dating the November 1962 publication of *Beat the Dealer* would establish with certainty that Dubner developed the Hi-Lo system without any external influence. However, the tapes were discarded by the family.

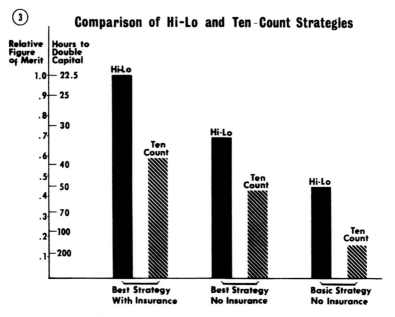

Dubner's comparison of his Hi-Lo system and Thorp's ten-count strategy. The image is taken from a photocopy of nine apparent slides most likely presented at the panel discussion on "Computers Applied to Games of Skill and Chance" at the 1963 Fall Joint Computer Conference. Little doubt remains that these materials, either as slides and/ or as handouts, were in fact presented at the panel discussion. This provides the results of the only known rigorous comparison of the two strategies. Thorp, on page 94 of the 1966 revised edition of *Beat the Dealer*, opines that the two strategies are of "comparable power." Dubner's analysis shows that the Hi-Lo is superior. The "Figure of Merit" is defined in the figure below.

Following the presentation, Thorp and Julian embarked on a detailed analysis of the system, Braun using the computer of the IBM facility in Chicago and the programming approach Thorp had used to develop the ten-count strategy he had introduced in the 1962 edition of *Beat the Dealer*. Thorp would refer to the system in the 1966 edition of *Beat the Dealer* as the "complete point count system." Some refer to it as the "high-low" system.

Gambling writer Jerry Patterson, in his column "Harvey Dubner: The Forgotten Man of Blackjack," suggests that without Dubner's development blackjack would not have become the most popular casino game. The ten-count strategy presented by Thorp in the 1962 edition of *Beat the Dealer*, of approximately equal power to the complete point system introduced in 1966, based on the Hi-Lo system of Dubner, required the player to calculate an index including the tenths and hundredths digits. This cumbersome necessity would severely limit the number of players able to play blackjack profitably. In Patterson's view, the development of the Hi-Lo by Dubner is responsible for the wide popularity of the field of card counting. Inexplicably, although my conversations with the Dubner family make it clear that his joy resulted from solving the problem, Dubner has not been elected to the Blackjack Hall of Fame. To me this is an egregious oversight. Without Dubner, no Blackjack Hall of Fame would exist.

In research, investigators build upon the work of their predecessors. Thorp was initially influenced by the work of Baldwin, Cantey, Maisel, and McDermott. Thorp's work subsequent to the 1963 Fall Joint Computer Conference was influenced by the work of Dubner. That one researcher builds upon the work of others is how scientific knowledge progresses. No one's work is diminished in importance as a result. As Sir Isaac Newton wrote (in a phrase first stated by the twelfth century scholar Bernard of Chartres), "If I have seen further it is by *standing on the shoulders of Giants.*"

Dubner was always and remains a quiet man, not interested in fame or fortune. His interest was mainly in the joy of solving mathematical and engineering problems. Like many of us interested in the mathematics of the game, he was not at all a gambler. It is appropriate that the lack of biographical material on his life, personal philosophy, research approach, and contribution to the game of blackjack be documented. That is done for the first time here.

The following is a moderately edited version of communications between myself and the Dubner family in January and February 2017. The first person (when I say "I," etc., refers to Dubner's son, Robert. Comments by his wife Harriet, who had a career as a speech pathologist, are enclosed in square brackets. Comments by me relating the content of phone conversations with the family or elaborating upon or clarifying concept are enclosed in curly brackets.

Professional Career

Dubner started working in 1949, just before graduating from college. His first jobs out of college were with the Arma Corporation (1949–1951) and with the Avion Corporation (1951–1960). Those jobs involved defense work, including fire control systems for U. S. Navy surface vessels at Arma and the design of the infrared seeker head for the Navy's Sidewinder missile, his main project at Avion.

The work at both Arma and Avion occasionally took him to the West Coast, and he developed a habit of stopping overnight in Las Vegas on his way home to New York. Because traveling from California to New York by turboprop aircraft was noisy, lengthy, and tiring, it was much more civilized to take a short hop to Las Vegas, spend the night there, and then take an early flight back to the East Coast than it was to travel by airplane at night. There, other engineers introduced him to Las Vegas and casino gambling.

It was during these visits to the casinos of Las Vegas that he developed his interest in possible advantage play in blackjack. It was obvious to him that knowing what cards had been played could lead to an advantage over the house, if only the task of recalling those cards and the resulting calculations could be made manageable. But it was after he started using computers in the late 1950s that he began thinking about using computers to develop a method that could be used by a gambler in a casino.

I became interested in my father's career and engineering in general when I was in the sixth grade, which was the 1964–1965 school year, my having been born in 1953. I was 11 years old that Christmas, and I remember a class project involving a mural depicting a winter scene. I augmented that mural by incorporating two NE1 neon bulbs as the eyes of the snowman. Powered by a 90-volt radio "B" battery, a simple RC driving circuit caused the snowman's eyes to blink.

(My teacher, who I realize in retrospect was a very sweet woman in her twenties, was utterly bewildered by me. I mean, c'mon. I was a scrawny 11-year-old kid, not even five feet tall, and I show up with a handful of electronic components and a 90-volt battery, connect them together with some alligator clip jumper cables – without a drawing – and the lights start blinking. She must've felt like she was falling down the rabbit hole.)

The key thing is that my dad brought those components home from his job at Simmonds Precision Products in Tarrytown, New York. The other key thing is that I was, at around that age, starting to pay attention to my father's work and career. I relate the anecdote about the snowman's eyes to establish my bona fides and to mark that point in time.

A few years later when I was in the ninth grade, the 1967–1968 school year, I learned to program computers. I worked with my father for the next 30 years, so I can lay some claim to being able to come up with some good guesses about what happened in the times before I really started paying attention.

Professionally, in the 1950s my father worked on missile guidance systems, in particular infrared homing systems. He was part of the team that developed the AIM-9 Sidewinder. He designed the optics, which meant ray tracing, which meant a number of people, mainly women, doing arduous calculations with mechanical calculators. That led him to learning the capabilities of electronic computers. He got hold of his first computer, a 1956 Royal-McBee LGP-30, in part to do those calculations.

He also became fascinated by catadioptric telescope design as a result of his optics work. In particular, he was much taken with the elegance of the Maksutov design.

In 1976, having been interested in astronomy and star-gazing all his life – when I was about 9 years old he got me an Edmund Scientific three-inch Newtonian as a Christmas present, and he messed with it at least as much as I did – and having lusted after one for years, he bought a Questar Standard 3.5-inch telescope. I have it now; I recently had it serviced and converted to the Duplex model, where it can be removed from its alt-azimuth mount and placed directly on a tripod for field work or as a telephoto lens.

[We moved to Ridgewood, New Jersey, in 1963, and he worked at Simmonds. I think he worked at Arma Corporation from 1949 to 1951 and then worked for Avion for 9 years. I believe he went to work at Curtiss-Wright in Paterson, New Jersey, in 1960 for a couple of years and then Simmonds also for a couple of years. In late 1963 he went to work at Computer Applications until that company went bankrupt, at which time he formed his own company, Dubner Computer Systems.]

I am pretty sure that he went to Computer Applications later than 1963. And I am pretty sure he brought those first components for me from Simmonds in 1964. He easily could have started the {blackjack computer simulation} work at Curtiss-Wright and kept it up at Simmonds. Or maybe he did it all at Simmonds.

Curtiss-Wright still exists, as does Simmonds Precision Products. It is an 85-year-old aviation company. The "Wright" goes back to the Wright brothers and the "Curtiss" goes back to Glenn Curtiss. Curtiss and the Wright brothers were vigorous competitors in the early days of aviation, but eventually their companies merged.

Although by the end of 1964 my father was at Simmonds, I am pretty sure that he was at Curtiss-Wright in the 1962–1963 time frame. We moved from Washington Township, New Jersey, to Ridgewood in June of 1963. Curtiss-Wright had many locations. I seem to recall that my father worked at the Curtiss-Wright location in Paterson. We used to go as a family to a restaurant named Steve's, which I believe was near where Dad worked. A dim memory has something on the wall of the restaurant relating to the Kennedy assassination, which took place in November of 1963.

Either Curtiss-Wright or Simmonds Precision Products would have had, in 1963, a print shop capable of creating those 3 × 5 cards.

The Computer Used in the Blackjack Simulations

I am pretty sure the computer my father used at Curtiss-Wright was the IBM 1620. The computer work was probably done off-hours at Curtiss-Wright. [I agree. It was at Curtiss Wright on off-hours.] My father had some big spools of black eight-track punched paper tape in his files. (I later worked with such tape for several years, so I have clear memories that it was the eight-track ASCII type, and not the earlier 6-track Baudot coded tape used in Flexowriters.) As I type this, I am having a visceral memory of how that heavy oily paper looked, felt, and smelled, as if were in my hand right now. I remember my father telling me that those tapes were the saved output of blackjack simulation runs. I am pretty sure the computer my father used at Curtiss-Wright was the IBM 1620. I can't swear to it, but the timing works. And the 1620 had the 1621 paper tape reader and the 1624 tape punch.

Although his memory is weak on this point, the work had to have been at Curtiss-Wright. He might have done some preliminary work at Avion on the LGP-30, but the LGP-30 used six-level punched tape. My memories are as clear as they can be after over 50 years, and although six-level and eight-level tape are each one inch in width, the hole patterns are different, and dang it all, the blackjack results were punched on the eight-level tape used on the IBM 1620. And, again, there doesn't seem to have been enough time after starting at Simmonds for him to have done the necessary non-work-related research there.

I know Harvey worked with the 1620. I have no memory of him working with the IBM 704 or 709. That doesn't mean much because until I started programming in 1968 I didn't pay attention to the "which computer" question. I know his first machine was the Royal McBee LGP-30, and I know about the 1620. But I don't know much about other machines he used at work until I really got involved around 1971. Online articles describing the IBM 704/709 don't mention punched paper tape. Those machines apparently used magnetic tape along with card punchers and readers for data storage. And I "know" he punched the output of his blackjack simulations on paper tape. This all points to the 1620, and away from the 704/709.

Without being sure, my guess is that he did the blackjack simulations on an IBM 1620; the timing is about right. [We moved to Ridgewood in June 1963 soon after Harvey went to work for Simmonds. But I think he did his simulations when he worked for Curtiss-Wright.]

{Only mild uncertainty remains in the family memory whether the blackjack simulations were performed at Simmonds or Curtiss-Wright. That the family moved in June 1963, after which he began to work for Simmonds, and that the 1963 Las Vegas conference was held in the fall of 1963, allowing sufficient time to prepare the paper and submit it for presentation implies the work was performed before the move, at Curtiss-Wright.}

I don't believe he "worked at night" on the computer. He would have written the code at night at home. He would have gotten it working during the work day. We always mixed in extracurricular stuff in with the paying work;

it was fun, it kept things interesting and exciting, you learned things that way, and the paying work "always" got top priority, with the extra stuff filling in the cracks. Then he would do the long runs for statistically significant results overnight, gathering the results the next morning.

A "long run" for him, given the power of the computers of the time, was on the order of ten thousand hands.

[Harvey says he was not working with anybody when he created his Hi-Lo system. And no he did not have a software collaborator. He was not in contact with Julian Braun. He says he might have been in contact with him after the {1963 Las Vegas computer} conference, but not for any reason.]

Dubner's Approach to the Blackjack Problem

As to what happened during the development of his counting rules, my knowledge of my father and the way he liked to work tells me it went like this. He played blackjack from time to time in Las Vegas and got interested in the possibility of advantage play. He would not have attempted to find anybody else's work on the subject. He and I talked about blackjack a lot over the years, and he never mentioned anybody but Thorp, but he never indicated that he learned about counting from Thorp.

Based on my recent long conversation with my parents, I hadn't fully realized how much casino blackjack he'd been playing during the 1950s. {That his work was performed without knowledge of Thorp's 1962 book} makes sense. There just wasn't enough time between publication of Thorp's book in November 1962, and the November 1963 conference for my father to have developed the expertise to play casino blackjack, work out the math, write the programs, do the simulations, and develop what he referred to as "his method." He had to have been working on it before Thorp published.

My father, through the years, and during recent discussions inspired by your questions where we have walked through his memories, has always said that he never even knew who Thorp was until the 1963 conference. He and my mother are in agreement that he played blackjack in Las Vegas starting as early as 1950 or so, and that his interest in advantage play grew steadily throughout that decade, culminating in his developing his method through computer simulation after the computers became available. My belief now is that the work was being done in parallel by Thorp and my father; neither knew about the other.

{This independence of effort is confirmed by blackjack writer Peter Ruchman, who quoted Thorp as writing: "At the 1963 conference Harvey Dubner presented the. .. (high-low system), which he had thought up himself.}

Having gotten interested, he would have done a mathematical analysis. Taking the formal math as far as he could, he would then have started writing computer code. He would have investigated random number generators and created his own. He would have figured out how to use an RNG to shuffle a simulated deck of cards – which you may know isn't actually as straightforward as it sounds; there are a lot of ways of screwing up both random number generation and shuffling – and then he would have started playing simulated games.

I don't know the details of blackjack, but I believe there are some straightforward rules for when to ask for additional cards based on your hand and the dealer's upcards. And once you have rules of play, you can simulate multitudes of hands and experiment with methods of counting and calculating your edge.

To elaborate on this concept, in one of your questions, you wonder if "intuition" had something to do with it. I can assure you that intuition was involved in the process of developing the hard theory, but it was the hard theory and the results of simulations that led to his results. This is how mathematical and scientific research works: an investigator starts with a suspicion, an intuition, a hunch, maybe even a wish. But those feelings have to survive rigorous analysis. Otherwise they are just guesses.

I speak as one familiar with this kind of thing. I know how I would respond if I were to get interested the way I believe that my father got interested.

First, I would learn how blackjack is played. I would have learned the rules of the game, and I would have investigated optimal play in the case where I have no knowledge about the constitution of the deck. I am talking about things like, if I am holding two tens, it's kind of stupid to draw. And if I am holding a ten and a two, it's kind of stupid to stand. (At, least, I think that's the case; I am not a blackjack player.)

And I would have verified mathematically that when the deck is rich in high cards that I am more likely to win, and when the deck is rich in low cards I am more likely to lose.

Operating on that omniscient level I doubt any simulation is necessary; if you know exactly what cards are left in the deck, then the probability of winning or losing can be calculated precisely.

I believe that for a full deck, this has long been worked out and represents the "basic strategy," which is how to best play your hand against each possible dealer upcard. That procedure is also the foundation behind casino rules and payouts; it what makes blackjack the game it is. But all of those conclusions are based on averages; they would have assumed that you are playing against an infinite deck of cards.

Knowing my father, he would have built a blackjack engine and run it some thousands of times in order to verify the "basic strategy." He would have done it for two reasons: First, to verify the published basic strategy. Second, in order to test his blackjack engine. If for any reason his results differed from the published versions, he would know that he had to check his implementation of the random number generator and his "basic strategy" decision making. He wouldn't expect to discover the published techniques to be wrong, but you never know. And it's very easy to make implementation mistakes, so he wouldn't have moved forward on a belief that the published techniques were incorrect until he had really verified everything else. The basic principle is that it's necessary to make sure your initial conditions match what came before you.

Now we get into "informed play": If you know the composition of the remaining cards in the deck because you kept track of every card that was played, you can, in theory, calculate precisely the odds of winning your next hand. It's a combination, of course, of every possible hand the dealer can deal both you and himself, but it is in principle calculable.

That's where artistry and simulation come into play. Perfect knowledge is too costly for a human to work with; he wanted to come up with something simpler, something that could be done without equipment at a casino table. He would have started with various scenarios: What if I know exactly how many Aces there are left, and how many face cards, and how many twos, threes, and fours? Something like that. Then he would have started simulating: The computer would play thousands of games. He would have grouped each hand into buckets based on the strategy he was testing at that time. He would have kept track of how many times each bucket resulted in a win. I feel confident that he iterated through that kind of process until settling on "count the high cards, count the low cards, subtract the lows from the highs, and divide by the number of cards left."

So the process would have started with the conviction that an edge was possible even with "chunky" knowledge of the composition of the remaining deck, rather than precise knowledge. Informed intuition would have led to the creation of various scenarios. Simulation would allow each scenario to be tested, and the results of the tests would lead to one that seemed optimal: That is, it got "good enough" results without requiring superhuman memory or arithmetical abilities.

Independent Derivation of Optimum Bet Size Criterion

It was at that point that Harvey ran up against the question of given that you have an edge, how much do you bet? Too little, and your winnings grow slower than they need to. Too much, and you'll go broke if there is a run of hands against you.

From those assumptions, you can infer that there is some optimal amount to bet to maximize the rate at which you make money, given the magnitude of the edge and the amount of money in your wallet. That's what my father figured out, and he did the analysis and came up with the optimal amount to bet once the card counting rules gave him the size of the edge. He learned sometime later that he had independently reproduced the work of John Kelly. I am making educated inferences about some of this, but he told me he learned of the Kelly Criterion after he had recreated it himself.

The math at that point was incomplete. The simulations would have shown: "When the hi-lo index is 'this,' then the probability of winning is 'that.' The next question becomes: "If I am playing blackjack and my probability of winning the next hand is 52%, how much should I bet?"

Imagine a coin-flipping game: Tails means you win your bet, heads means the casino takes your money. You somehow know the coin is biased to come up tails 52% of the time, and you know that if you play long enough you'll come out ahead. How much should you bet?

If you have $100 and place it all on the table, you have a 48% chance of going broke on the first flip. Bet nothing and you might as well have stayed home.

②   **Return on Investment Theory**

1) $\text{PROBABILITY OF RUIN} = P_{(o)} \overset{*}{=} \left(\dfrac{q}{p}\right)^{\frac{\text{TOTAL CAPITAL}}{\text{BET PER TRIAL}}}$

2) $\qquad P_{(o)} \approx e^{-\left(\dfrac{2 \times \text{EDGE} \times \text{TOTAL CAPITAL}}{\text{BET PER TRIAL}}\right)}$

3) $\dfrac{\text{EXPECTATION PER TRIAL}}{\text{TOTAL CAPITAL}} = \left(\dfrac{\text{EXPECTATION PER TRIAL}}{\text{STANDARD DEVIATION}}\right)^2 \times \left(\dfrac{2}{-\text{LOG } P_{(o)}}\right)$

4) $\text{FIGURE OF MERIT} = \left(\dfrac{\text{EXPECTATION PER TRIAL}}{\text{STANDARD DEVIATION}}\right)^2$

5) $\text{RETURN ON INVESTMENT} = \left(\text{FIGURE OF MERIT}\right) \times \left(\dfrac{2}{-\text{LOG } P_{(o)}}\right)$

* p = probability of success
  q = probability of failure

Dubner's summary of his consideration of betting in terms of financial theory. The image is taken from a photocopy of nine apparent slides most likely presented at the panel discussion on "Computers Applied to Games of Skill and Chance" at the 1963 Fall Joint Computer Conference. Little doubt remains that these materials, either as slides and/or as handouts, were in fact presented at the panel discussion.

The appropriate manner in which to analyze this game is to calculate the expectation value per flip. Because you would have a 52% chance of winning $0.01 and a 48% chance of losing $0.01, that would be $E = 0.52 \times \$0.01 - 0.48 \times \$0.01 = \$0.0052 - \$0.0048 = \$0.0004$ for the penny bet. Normally, expectation values are presented in terms of a $1 bet, in which case it would be presented as $0.04. You have a $0.0004/$0.01 × 100 = 4% advantage against the house. If you played 3600 flips, your expected winnings would be $3600 \times \$0.0004 = \$1.44$.

So, given a known edge in the game, there clearly is some point between betting nothing and betting everything you have that will maximize your winnings per hand over many hands. That can be plotted and analyzed as winnings-per-hand against size-of-bet (with both axes expressed as a fraction of your stake); it is a downwardly convex curve with the optimum at a peak.

That point can be analytically found using simple calculus. Finding good numerical values would be trivial using computer simulation. I asked my dad about it once, and he told me that he had done the math. This is exactly what John Kelly did and published in 1956 and which became known as the Kelly Criterion, or a Kelly Bet. My father learned about Kelly's work later.

So, there it is. Some of that I know my father did because he told me: He independently duplicated Kelly's math, and I know he simulated tens of thousands of blackjack hands. What I don't know is precisely what he was testing in those simulations. I do know the results he arrived at. And I worked with him often enough on other similar investigations to be morally convinced that he arrived at the hi-lo index by testing a number of other, similar, ways of counting the cards, and keeping track of which method seemed to work best in terms of being feasible at a casino table and effective in generating returns.

Dubner Worked Independently of Others on Blackjack

I can't comment on any possible competition in the development of the theory of card counting, or who might have appropriated work from somebody else. I am certain Harvey did his work independently. My understanding is that Mr. Thorp, in the later edition of *Beat the Dealer*, acknowledged Harvey's work and credited him with putting card counting onto a solid mathematical footing.

[Harvey is sure he never heard of Thorp until the 1963 joint computer conference. He is sure he worked independently. Harvey said he created the {3 × 5} card before he even knew who Thorp was.]

You ask about the term "basic" on the Hi-Lo card. I have no direct knowledge of why he used it. But knowing my father, I am positive the word means, "There is more to the story." As I mentioned earlier, the method on the card wasn't the best possible method. It was my father's attempt to balance practicality with effectiveness. I very strongly doubt that he even knew of anybody else's work, much less was taking anybody else's work into account. I don't know if he had published any papers at that point in his career, but he shortly would start to do so. If anybody else's work had informed his, I am certain he would have credited them. And I'm pretty sure it would have come up in our conversations over the years.

I figure also that card counting research and development in the early 1960's was a case of "when it's time to railroad, railroads pop up all over." I've been told that various gamblers had been trying to count cards for years. They kept quiet about it, so as not to lose what little edge they had, but word was spreading. J. L. Kelly published his work on optimal betting in 1956. Computers were becoming more generally available. So it's not surprising to me that a lot of activity was happening in parallel.

Dubner's Personal Interest in Blackjack

[He says what started out as an interesting mathematical problem became a big thing. And as time went on, it became bigger and bigger. At the beginning, he had no idea of the ramifications of his discovery. Harvey says suddenly he was involved in a situation where he was making money and had a good chance of making a lot of money sometime in the future. But he was doing other things professionally that were even more interesting. He was doing more and more work that was much more significant than card counting. Remember always he was a mathematician, not a gambler. He never benefited financially from his blackjack creations. He was a mathematician, never a gambler.]

I will reiterate the point: My father wasn't a gambler, and he had no particular interest in blackjack. But having learned that it might be possible to get an edge on the casino by tracking what cards had been played and thus developing some knowledge of what was left in the deck, and having his interest piqued by that, he would have – based on many discussions about this over the years – done the analysis, done the math, written the code, done the simulations, and formulated his method, completely on his own.

For complex problems, especially work related, sure, he would research, and collaborate, and build on the work of others. He did that a lot. But he was doing the blackjack thing for fun.

He was proud of how easy his method was to use. Perfect counting could do a little better than his rules, but it required perfect memory and a complex series of responses, and it could only do a few percent better. At least, that's what he told me.

So he twisted the tail of the casinos with that 1963 paper. I am absolutely sure he did it on purpose; his sense of humor is extremely puckish.

Family Claim that Dubner Was Not Aware of Thorp's Early Work

Based on my recent long conversation with my parents, I hadn't fully realized how much casino blackjack he'd been playing during the 1950s. I hadn't realized just how much blackjack he played in Las Vegas during the 1950s. My father, through the years, and during recent discussions inspired by your questions where we have walked through his memories, has always said that he never even knew who Thorp was until the 1963 conference. We've been talking to him about Thorp, and he says definitively that he didn't know about Thorp's work prior to 1963. He and my mother are in agreement that he played blackjack in Las Vegas starting as early as 1950 or so, and that his interest in advantage play grew steadily throughout that decade, culminating in him developing his method through computer simulation after the computers became available. My belief now is that the work was being done in parallel by Thorp and my father; neither knew about the other.

{The independence of effort in developing the Hi-Lo strategy is confirmed by blackjack writer Peter Ruchman, who quoted Thorp as writing: "At the 1963 conference Harvey Dubner presented the. .. (high-low system), which he had thought up himself."}

As I noted, his work taking him to Las Vegas in the early 1950s led him to start playing blackjack. Being mathematically inclined, he would have quickly figured out the theoretical possibility of advantage play. After he started using computers in 1956, he would have automatically started thinking about using them to develop workable techniques. He remembers doing just that.

One thing my father is adamant about: He worked out the possibilities of what became his hi-lo strategy on his own. It grew out of playing casino blackjack from time to time during the 1950s and the early 1960s. I have repeatedly revisited with him the question of "Where did you get the idea that counting the cards could lead to an edge over the house?" His answer is unwavering: "I worked it out myself." When asked if he learned of the possibility that he learned of it from Edward Thorp, his unwavering response: "I never even heard of Thorp."

{That his work was at least begun without knowledge of Thorp's 1962 book} makes sense. There just wasn't enough time between publication of Thorp's book in November, 1962, and the November, 1963, conference for my father to have developed the expertise to play casino blackjack, work out the math, write the programs, do the simulations, and develop what he referred to as "his method." He had to have been working on it before Thorp published.

If nothing else, he wouldn't have been in a casino often enough. And this was all extracurricular work. He wouldn't have let it interfere with his job, so his time commitment wouldn't have been great. I had previously thought that he learned about it from Thorp but that can't be right; there just isn't enough room in the timeline. And when I was doing that speculation, I didn't know he'd been regularly going to Las Vegas during the 1950s.

I figure also that card counting research and development in the early 1960s was a case of "when it's time to railroad, railroads pop up all over." I've been told that various gamblers had been trying to count cards for years. They kept quiet about it, so as not to lose what little edge they had, but word was spreading. J.L. Kelly published his work on optimal betting in 1956. Computers were becoming more generally available. So it's not surprising to me that a lot of activity was happening in parallel.

[Harvey says he was not working with anybody when he created his Hi-Lo system. And no he did not have a software collaborator. He was not in contact with Julian Braun. He says he might have been in contact with him after the {1963 Las Vegas computer} conference, but not for any reason. Harvey is sure he never heard of Thorp until the 1963 joint computer conference. He is sure he worked independently. Harvey said he created the {3 × 5} card before he even knew who Thorp was.]

So, putting together everything I know now, it's clear to me that Harvey was one of a small number of people at that time who had the interest, the inclination, the mathematical training, the innovative insight, the computer programming ability, and — not insignificantly in that time frame — the access to a computer necessary to conceive and build the simulations needed to create and test useable blackjack counting strategies. But it is clear to me that he started his work because of the possibilities that became apparent to him while playing and thinking about blackjack, and not because he learned about card counting from anybody else.

Concerning the comparison of the Hi-Lo and Ten-Count strategies {displayed above}, does it demonstrate that Harvey knew about Thorp and the Ten-Count Strategy at the time of the conference? Sure does. That leaves the question you find most interesting: Does it show that Harvey learned about counting from Thorp? Harvey is adamant that he developed the idea on his own. My belief, knowing him, is that he did develop it on his own. My belief is that when he learned of the Ten-Count strategy he compared it to his method, found his to be superior, and that may well have led to him deciding to present the Hi-Lo strategy at the Fall 1963 conference. {Or, he may have already decided to present at the conference before reading *Beat the Dealer* and simply added the comparison to the material he had already planned on presenting.}

Once he learned of the ten-count strategy. he would very naturally have compared it to his. And upon finding his Hi-Lo strategy to be superior, and then learning of the panel session at the Fall Joint Computer Conference, it's natural that he would have used it as an opportunity to go public. It's distinctly possible that he would have gone to that conference even without the "games of chance" session; he was a computer professional at a time when it was still a new field. I note that he had a paper published at the 1970 Spring Joint Computer Conference (http://dl.acm.org/citation. cfm?id = 1,476,964). I am speculating that he was planning on that trip to Las Vegas anyway, and the panel discussion was a chance to have some fun and show off his work.

It seems clear that by mid-1963, my father had read the 1962 *Beat the Dealer*, because it appears from that handout {Fig. 3} that he tested Thorp's ten-count system using the same simulation techniques he used for developing his Hi-Lo. {The format employed by Dubner in the handout does not allow immediate comparison with Thorp's published results, and this comparison would likely have required the additional simulation to which Robert Dubner refers.} But there is no question in my mind that he did the bulk of his work before he read *Beat the Dealer*. He started his work and he developed his methods, and then read *Beat the Dealer*. Remember, he wasn't a professional gambler. His interest was theoretical and not because he wanted to win at blackjack. He wouldn't have bought the book *except* for the fact that he was already doing research and development on card counting.

How much of his final work was done after reading the book versus before nobody will ever know. My guess is "not much;" the essence of my dad's method is that it is *simple*; it would have coalesced fairly early.

{The 1962 copy in their family library is the hardcover first edition from Blaisdell Publishing Company.} I see one rare book dealer offering one in fine condition for $500. My mother later commented that maybe we should sell it.

{The findings of the family of the copies of apparent slides which may also have been used as handouts at the 1963 conference led to a further consideration.} There are a number of interesting things in that handout. He apparently did some simulations that involved hundreds of thousands, even millions of hands. Pure speculation, of the type that I really should stop making: I know he has told me in the past that his runs involved tens of thousands of hands. This leads me to wonder if he started his work on the slower Librascope LGP-30 {while he worked at Avion, from 1951 to 1960}, and later refined it on the faster IBM 1620 {at Curtiss-Wright}. But that's absolutely pure speculation. {If true, Dubner clearly began his investigations before the publication of the 1962 first edition of *Beat the Dealer*.} The letter-sized sheet of paper was prepared at the same time as the 3x5 card, which means at some point in late 1962 or early 1963 when he still had access to the 1620 computers at Curtiss-Wright.

If Thorp's account in the 1966 edition of Beat the Dealer is accurate, Harvey didn't present those sheets at the meeting. Thorp says on page 94 of the paperback 1966 edition {of *Beat the Dealer*} that "Exactly how much better or worse (the Hi-Lo system) is than the Ten-count method is not known." It sure doesn't sound like Thorp saw those sheets of paper. My dad was a natural showman; he may have decided that the sheets were too technical and he should just stick with the "We're in Las Vegas, and with this little card you can beat the house!" fireworks, and leave presenting the math for another time…a time that never came. {I'd be shocked if Harvey didn't present the sheets, as handouts or slides/handouts. This was a technical meeting and he wouldn't just provide the slide of the 3 × 5 card. Thorp, as stated, remembers that the material was presented. His comment in *Beat the Dealer* may resulted from either not at the time of its writing remembering the details of Dubner's presentation or not wishing to quote results from a study the details of which he had not examined or verified for himself.}

I can't comment on any possible competition in the development of the theory of card counting, or who might have appropriated work from somebody else. I am certain Harvey did his work independently. My understanding is that Mr. Thorp, in the later edition of *Beat The Dealer*, acknowledged Harvey's work and credited him with putting card counting onto a solid mathematical footing.

I very strongly doubt that he even knew of anybody else's work, much less was taking anybody else's work into account. I don't know if he had published any papers at that point in his career, but he shortly would start to do so. If anybody else's work had informed his, I am certain he would have credited them. And I'm pretty sure it would have come up in our conversations over the years.

For complex problems, especially work related, sure, he would research, and collaborate, and build on the work of others. He did that a lot. But he was doing the blackjack thing for fun. He paid a lot of attention to blackjack over the years, and many of his memories are pretty clear. He doesn't remember where he did the work, but he does remember doing it, and he remembers why he did, which was for the fun of it and to prove that you could get an edge over the house in casino blackjack.

He wouldn't have found it to be fun to start with somebody else's work. It would have been far more characteristic of him to start from scratch. He was not surprised to learn that he was among a group of people doing similar work. To some extent it was the availability of computers that led both Thorp and my father to do the work they did. When it comes time to railroad, railroads pop up all over the place.

In addition to his current assertions, I had many conversations with dad about blackjack and counting over the years. He never mentioned Thorp except in the context of what happened after the 1963 conference. He never once said that he learned about the possibility of getting an edge from anybody else. He always spoke of it in terms of figuring out that it could be done and working out how on his own. He never spoke about building on anybody else's work, or working with anybody. He always spoke about working out the possibility himself, and working the math and the computer simulations to come up with his method. He never referred to it as anything but "his method." For example, he would talk about how "theoretically there are better methods than 'my method,' but they are much harder to use, and the gains are small." He didn't even use the name "Hi-Lo" when talking to me, and he didn't even call it "card counting" much, except when talking to other people. When he and I discussed it, it was always just, "my method."

I have always been interested in the technical aspects of his work on blackjack, and we talked about it many, many times. He always — always! — spoke in terms of what he had figured out. He never once said that he had learned about the possibilities of card counting from somebody else.

And if he had, he would have acknowledged it. I worked with him for years doing number theory research. He learned a lot from other investigators, like Richard Brent, and Peter Montgomery, and Hugh Williams, to name just a few. He often spoke in terms of what he had learned from those other people. He was always careful about what he learned from other people, and the original work he did himself.

If the idea that Harvey did the work all on his own is controversial, well, it's only because nobody ever asked him. He's been in the phone book all this time, and for many years his e-mail address has the same and available. If anybody asked, as you are asking, how, he would have said, as he is saying now, that he did it all on his own.

If somebody can demonstrate that they were in touch with my father about blackjack before he gave that presentation, I would be both astonished and incredibly interested in talking to them.

If somebody thinks they can prove my father read Thorp's book {before he began his own investigations}, or somehow got the idea for card counting other than on his own, well, I'm pretty sure that his reaction over the years would have been to ignore them. He was always like that. He knew what he'd done and, although he liked being acknowledged for it, he didn't go out of his way to correct misapprehensions. Heck, I happened to pick up a copy of {William Poundstone's} *Fortune's Formula* a few days ago. Lo and behold, my father's counting method is described in there, but without attribution, except vaguely in Thorp's direction.

So it's not surprising that dad's method evolved into being described as "an improvement over Thorp's method."

It's incontrovertible that Thorp did publish before my father gave his presentation in 1963, so it's perfectly reasonable for the world to have concluded that Dubner's method was an improvement over Thorp's. But Harvey is clear that he didn't know about Thorp's work, and his doing the work *de novo* fits every discussion he and I ever had about it. {At this point, after discovery of the apparent slide comparing the Hi-Lo strategy to Thorp's ten-count strategy, Robert Dubner would state that his father did not know about Thorp's work at least when he began his investigations and possibly until most of it was completed.}

Presentation of Hi-Lo at the 1963 Fall Joint Computer Conference

According to the program as published in the {September, 1963, issue of the} trade journal *Electrical Engineering* the conference was to take place in the Las Vegas Convention Center. You'll note at the bottom right of page 20A in the program {see bibliography} that on Wednesday, November 13, at 8:00 p.m., "Computers Applied to Games of Skill and Chance" was scheduled, chaired, inexplicably, by "R.A. Kudlich, General Motors Corp."

In the "backmatter" {of the published proceedings; see bibliography}, "E. O. Thorp" is listed as one of the Session Chairmen. {This information, however, was published in the proceedings of the conference, and not before it. Dubner first encountered Thorp at this conference, according to his recollection.} In the second big box on the page {see bibliography} there is a set of tabs, labeled Abstract, Source Materials, Authors. . ... The Source Materials tab has a link to the Back Matter.

{The proceedings were published both by the Association of Computing Machinery and Spartan Books of Baltimore, Maryland. Neither, however, provides excerpts from the games of skill and chance panel discussion, only papers formally presented to the various conference sessions. The 3 x 5 index card on which Dubner printed the "Basic HI-LO Strategy" and, if they were presented, the nine slide/handout material, are the few, if only, artifacts remaining from the panel discussion on games of skill and chance.}

Dr. Kudlich was a long-standing and reliable administrator of the IEEE, and he apparently organized the 1963 "Computers in Games..." panel. It seems equally clear to me that Thorp was brought in as the pre-eminent popular figure, because of *Beat the Dealer*, to chair the actual session. (An inspired move, if you ask me!) My educated guess is that Kudlich was probably there, probably introduced Thorp to the enthusiastic crowd, and then Thorp moderated the discussion. In short, it really looks to me like Thorp was rightly brought in as a celebrity figure to moderate the panel. {At the date of this writing, Thorp has not been able to provide information from his files as to when he was invited to moderate the panel, when and how that development was publicized, if he selected the members of the panel, and if Kudlich in fact introduced him.}

I asked him about how he got to present at the 1963 conference. He said that he learned about the panel discussion on computers and games of chance — the topics are published in advance — and he contacted the conference and asked for a slot. {That is, his participation was not based on personally knowing Thorp.} (My father is a lifetime member of the IEEE, and *Electrical Engineering Magazine* was an organ of the IEEE.) And he simply doesn't remember. The answer {to exactly how he learned of the panel discussion}, again, is that nobody knows; nobody can know. But you can be absolutely sure that Curtiss-Wright had a library that subscribed to all the publications of all those groups, as well as the publications of the American Federation of Information Processing Societies. At that time a company like Curtiss-Wright would have had all of them available, and my father and his coworkers would have paid attention to them. The astonishing thing would have been if a panel on "Computers Applied to Games of Skill and Chance" had gone unnoticed by my father, given his interest at that time in blackjack. I don't know who R. A. Kudlich was, but it's easy to think that when that September 1963 magazine was put to bed Edward Thorp hadn't yet agreed to chair the panel.

Casino Reaction, Dubner's Personal Blackjack Winnings, and His Personal Reflections

I know people who have made tens of thousands of dollars using his method. Hell, I know a guy who claimed he bought a house with card-counting winnings. But not my father. As my mother and I have said, he wasn't a gambler. He is, in fact, remarkably risk averse. He has told me that when playing at a one dollar table when the deck got hot, his hand would shake when he put down the twenty dollar bet.

The pickings were easy in the beginning. The casinos didn't know what was going on, and particularly downtown in Vegas the dealers would play a single or double deck right down to the bottom.

As time went on, of course, you know what happened. Two, four, even six decks. The yellow "stop here and shuffle" card halfway down. Lately the casinos, I am told, have deployed their Doomsday Machine: continuous mechanical shuffling, wiping out any hope of an edge.

And, of course, the dealers themselves had nothing better to do than count the cards, and when they saw the counting betting pattern they'd signal the pit boss, who'd intervene. That led to the team play concept made noisily public by the MIT crew, where one person would quietly count and signal a cowboy to come play a few hands when the deck got hot. But you know all about that.

People have a tendency to gleefully assume that my father was banned from casinos all over town. Not so. Nobody knew who he was by sight. He kept his head low when he was playing blackjack. The one and only time my father was asked to leave — it was at a downtown casino, I think he once told me — he was actually behind at the time. "But I'm losing!" he told me he told the pit boss. The polite reply was something like, "We don't want your money."

[He never stayed long at any casino. He didn't want them to know he was counting. Even so, at one point we were asked to leave although he was betting very little. I don't remember the name of the casino. We stayed at the Riviera so I guess we played more there than anywhere else. We also played in town.]

My father wasn't there to win. He was there for the satisfaction and fun of experiencing his results at work. He kept careful records; I am sure you will start slavering when I tell you those records still exist, as far as I know. Among other things there is a graph, covering at least a decade or two of once or twice a year visits to Las Vegas, showing the amount of money earned per hour of play. It slopes steadily down to the right, both because of some outsized returns early on — a case of "beginner's luck" — combined with the effects of the steadily escalating countermeasures by the casinos over the years.

[He says what started out as an interesting mathematical problem became a big thing. And as time went on, it became bigger and bigger. At the beginning, he had no idea of the ramifications of his discovery. Harvey says suddenly he was involved in a situation where he was making money and had a good chance of making a lot of money some time in the future. But he was doing other things professionally that were even more interesting. He was doing more and more work that was much more significant than card counting. Remember always he was a mathematician, not a gambler. He never benefited financially from his blackjack creations. He was a mathematician, never a gambler.]

I will reiterate the point: My father wasn't a gambler, and he had no particular interest in blackjack. But having learned that it might be possible to get an edge on the casino by tracking what cards had been played and thus developing some knowledge of what was left in the deck, and having his interest piqued by that, he would have — based on many discussions about this over the years — done the analysis, done the math, written the code, done the simulations, and formulated his method, completely on his own.

He was proud of how easy his method was to use. Perfect counting could do a little better than his rules, but it required perfect memory and a complex series of responses, and it could only do a few percent better. At least, that's what he told me.

So he twisted the tail of the casinos with that 1963 paper. I am absolutely sure he did it on purpose; his sense of humor is extremely puckish.

{He did make enough money on one trip to buy his wife a $200 diamond pin, which they still possess.} [We bought the pin in Ridgewood, New Jersey. I believe the name was Webber Jewelers. That's just a vague memory.] The jeweler in Ridgewood was Weber Jewelers.

You wonder if those sheets of paper {containing the nine materials} were created after 1963 for a second presentation. I tell you that they were not. There is no evidence anywhere that my father ever talked publicly about blackjack counting ever again. He talked about it plenty privately. He made no secret about it. He relished having created the first practical blackjack counting system, and would tell anybody about it at the drop of a hat. But he wasn't part of the gambling world, and didn't expend any energy pursuing it. He never did any subsequent research or simulations.

After 1963 he was done, except to go to Las Vegas and occasionally Atlantic City and count at blackjack, keeping track of his winnings. His long-maintained gently downward sloping graph of $/hour over time appears to be lost, but he was still winning when, around 1977, not long after the sole time he was asked to leave a casino because he was caught counting, he gave it up. I asked why he stopped. He said it was because it had become boring.

But my father had a unique opportunity. A mathematician at heart, he liked to prove things formally. As an engineer, he would switch to pragmatic methods when the formal math fell short. Having embraced computers for solving problems starting in 1956, he knew when to switch from formal methods to computational ones. He had the computer programming skills to implement the methods, and he had access to the computer needed to perform the calculations. He also had the wry sense of humor needed to publish his results in Las Vegas.

About the Author

Son, someday you will make a girl very happy, for a short period of time. Then she'll leave you and be with new men who are ten times better than you could ever hope to be. These men are called "astronomers."

Leslie M. Golden is a Renaissance Man. Not only a multiply-awarded writer of fiction, non-fiction, and commentary, he is an astronomer, professor, musician, bandleader, actor, stand-up comic, software developer, published editorial cartoonist, cruise ship lecturer, and animal welfare and environmental activist.

Chris Lines, the editor of *Gambling Online* print magazine writes, "Where to start in introducing the complete one-off individual that is Les Golden? Actor, stand-up comedian, humorist, UFO lecturer, singer, astronomer, cartoonist,

© Springer International Publishing AG 2017
L.M. Golden, *Never Split Tens!*, DOI 10.1007/978-3-319-63486-9

playwright, trumpet player, voiceover artist, political activist… we could go on. You can probably tell that Les is a bit of a character. Luckily for readers, he's also a great blackjack player."

Eric Zorn of the *Chicago Tribune* in a column devoted to Les Golden describes him as "an unusual man. He is an actor and educational software developer with a Ph.D. in astronomy; he is a trumpet player, writer and physics professor."

Compuserve Magazine notes that "his interests form a list so long as to stagger the imagination. He is a stand-up comic who has performed all over the United States and Mexico, a professional actor in more than 100 plays, films and commercials; and he is the author of *Basic Composer*, software that is used to compose, play back, and print music and lyrics."

Dave Bland, the editor of *Flush Magazine* and a British television pundit provides a succinct perspective: "Les Golden is a comedy genius. It really is as simple as that."

His wide range of interests led to a long-running Chicago-area newspaper comic strip to feature him as "Moe Silver" (pun on "Les Golden").

Leslie Morris Golden received the B.A. with Distinction and Masters in Engineering Physics from Cornell University, where he was a McMullen Fellow and a Fellow of the Interfoundation Committee of the American Institute for Economic Research (Great Barrington, Massachusetts). He performed his masters' thesis on erosion of lunar craters by micrometeorite impacts under Bruce Hapke and studied radio astronomy under his early mentor Frank Drake of SETI fame while working summers performing planetary exploration simulation studies at the AstroSciences Center of the Illinois Institute of Technology. He was the award-winning editor-in-chief of the *Cornell Engineer* magazine and was a trumpet soloist with the Cornell Concert Band.

He received the M.A. and Ph.D. in astronomy from the University of California, Berkeley, where he performed his dissertation, "A Microwave Interferometric Study of The Subsurface of the Planet Mercury," under William J. ("Jack") Welch, the Watson and Marilyn Alberts Professor of Extraterrestrial Intelligence. It was cited as one of only two dissertations for excellence in the year of completion by the Berkeley chapter of Phi Beta Kappa. The models he developed form the basis for many thermophysical models of planetary surfaces currently employed.

He obtained a National Research Council Resident Research Associateship grant to continue studies of Mercury at the Jet Propulsion Laboratory and studied the time variability of quasars at the Aerospace Corporation before obtaining his position as a professor at the University of Illinois at Chicago, where he became affiliated with both the physics department and the Honors College. He performed additional study on fellowship at the Kellogg Graduate School of Management Northwestern University and has taught mathematical methods and statistics and probability in the Heller Graduate School of Business of Roosevelt University in Chicago.

His research and scholarly interests are planetary radio astronomy, observational selection effects, cosmology, the possibility of extraterrestrial life, and the history of astronomy. He has published numerous technical articles in his field as well as the textbook *Laboratory Experiments in Physics for Modern Astronomy*. He is one of a handful of Americans to be elected to both Tau Beta Pi, the engineering honorary society, and Phi Beta Kappa, the arts and sciences honorary society. He was also elected to Pi Delta Epsilon, the journalism honorary society. He is listed in Marquis Who's Who in Science and Technology and Marquis Who's Who in the World.

In demand as an authoritative and entertaining public speaker and lecturer, among his many appearances he was selected by Royal Cruise Lines to be their on-board lecturer on the high seas on the S.S. Royal Odyssey during the 1986 apparition of Halley's Comet, was the featured keynote speaker on the occasion of the dedication of the new wing of the Adler Planetarium in Chicago at its annual meeting, and provided a series of lectures on extraterrestrial life at the Field Museum of Natural History. In 1996 he was the first University of Illinois at Chicago faculty member to be selected a visiting professor on the Semester at Sea program of the University of Pittsburgh, traveling around the world teaching undergraduate courses on astronomy and the possibility of extraterrestrial life. He speaks widely to library, school, youth, and adult groups on various topics in astronomy, particularly the possibility and nature of extraterrestrial life, and fulfillment of one's potential.

He was a finalist to be a member of the second crew of Biosphere II in Oracle, Arizona, before the changing of its mission from scientific experimentation performed by isolated Earthlings to public education. From there he was to direct the Near Earth Asteroid Reconnaissance Project (NEAR), his search coordinating amateur astronomers from around the world to search the heavens for asteroids on Earth-collision path.

Les has additional careers as a professional trumpet player and vocalist, bandleader, stand-up comedian, both as himself and as Bhutanese comedian Subrahmanyan Berkowitz, and professional actor, mentored by improvisation guru Del Close of Second City, and cartoonist. He studied trumpet with Ben Purdom, Jerold Cimera, and Adolph "Bud" Herseth and has performed with and led dance bands since high school.

At the University of California, he was one of the founding fathers, a trumpet player and vocalist, and the long-time announcer for the prestigious University of California Jazz Ensembles as well as a popular emcee for jazz festivals throughout California. He is an accomplished scat singer and has performed as a jazz vocalist nationwide and in Mexico. He formed and led the Les Morris Quintet. His best-selling computer software program, *Basic Composer*, allows musicians to compose, transpose, and print out music and lyrics.

As "Flash Golden," he originated and hosted Flash's Jazz Patio for 7 years in the San Francisco Bay Area as a disc jockey on KALX-FM, and was the play-by-play voice for California Golden Bears basketball.

He performs as a stand-up comic and in theatre, television, radio, and motion pictures. As a stand-up comedian, he has performed nationwide, at the Holy City Zoo in San Francisco, the Comedy Store in Los Angeles, the Comedy Cottage in Chicago, as well as on the college circuit, Playboy Club, resorts, other comedy clubs, and on television and radio. He appears both as himself, Les Golden, and as Subrahmanyan Berkowitz from Bhutan.

Les Golden has appeared in over 100 stage, motion picture, radio, television, television commercial productions, and magazine and newspaper ads. He has appeared numerous times as an actor on the live radio broadcast productions of Unshackled! and was a frequent regularly featured comic guest on the Eddie Hubbard Show radio program broadcast live from Arnie's restaurant in Chicago as the comic character Jeffrey Clayton Maxwell from Bhutan. He was a charter member of the Chicago-based Porchlight Theatre Ensemble and has appeared in film and television in featured roles with Forrest Tucker, Tippi Hedren, Troy Donahue, Charlotte Ross, and others. His stage credits range from a comic Indian guru to Shakespeare. Les is a member of both the Screen Actors Guild (SAG) and the American Federation of Television and Radio Artists (AFTRA). His Bacon number (the number of motion picture role links between a given actor and prolific actor Kevin Bacon — Les Golden, *Deadly Spygames* with Tippi Hedren, who was in *Jayne Mansfield's Car* with Kevin Bacon) is 2 (closer value than 82% of all credited motion picture actors).

As a writer, Les has published in various genre including screenplays, theatrical plays, sitcom scripts, a political candidate guide, astronomy textbooks, gambling, humor and as a jazz critic.

He is internationally known as a popular and authoritative yet entertaining gambling writer, having written over 100 columns applying probability and statistics to blackjack, craps, and roulette for five London-based gambling magazines. As detailed in series of columns, he developed the Golden Diagram system for winning at blackjack and the Magic Circle strategy for winning at roulette. The website of *Gambling.com* magazine refers to him as "gambling.com magazine's resident blackjack genius." In addition to numerous technical articles in astronomy, his mathematical-based researches into the gambling games of blackjack and roulette have been published in peer-reviewed scholarly journals. He was named a 2016 Scholar by the International Gaming Institute of the University of Nevada, Las Vegas.

He mastered blackjack card counting systems while a graduate student at the University of California, Berkeley, and has used that skill from Las Vegas,

Reno, Lake Tahoe, and Carson City to Atlantic City, from Monaco to Macao, and on board cruise ships as a professional card counter.

He has won numerous awards for both fiction and non-fiction writing. These include the prestigious Eric Hoffer and Lili Fabilli Laconic Essay Prize, the International *Compuserve Magazine* Essay Contest, an award in the Griffith Observatory Science Writing Contest, and First Prize in the Senior Division of the Nicolaus Copernicus International Essay Competition commemorating the anniversary of Copernicus' 500th birthday (American Council of Polish Cultural Clubs).

Les applies his comic skills to cartooning and has had numerous cartoons published in newspapers and magazines. He is listed in the international Comiclopedia database.

He has been a lifelong avid baseball player, leading the league in hitting and batting 4th on the all-star team during his youth in Oak Park, and then managing and playing third base with his teams, his Cornell University Tau Epsilon Phi fraternity, the "Goldenrods" as a graduate student at Cornell, the "Foul Balls" as a graduate student at Berkeley, the championship Jet Propulsion Laboratory fast pitch team in the Glendale City League, and with the Emilia Lorence Talent Agency team in the Chicago Theatre League.

He is a fervent environmentalist and passionate and nationally-known animal welfare advocate, working nationwide to save dogs, abused circus and zoo animal inmates, and other animals and to reform animal laws.

His other interests include running, cooking, gardening, and photography. He lives in Oak Park, Illinois, in the Golden family home, home to Golden Gardens Pee Gee Ball Field, which he shares with his rescued canine and feline friends and occasional injured or orphaned wildlife.

The author (*left*) and his devoted and loyal pal, Cicero (*ball in mouth*). Rescued from the streets, my beautiful, athletic, intelligent, spirited, protective, obedient, always wanting to please, and bonded best friend.

* * *

We provide a few webpages about Les Golden and the film version of *Never Split Tens.*

Details about Les Golden's Renaissance Man activities can be found at http://encyc.org/wiki/Les_Golden

The finances of producing *Never Split Tens* as a motion picture are presented at http://encyc.org/wiki/Never_Split_Tens

The complete formal prospectus for the film is provided at http://en.wikisage.org/wiki/Never_Split_Tens_(movie_prospectus)

Les Golden's activities as a gambling writer are profiled on his *alma mater* website, http://www.cornell65.com/interest/golden.html

Les' listing on the Comiclopedia International Database appears at http://www.lambiek.com/artists/g/golden_les.htm

The *Chicago Tribune* presented a political profile of Les Golden at http://articles.chicagotribune.com/1995-10-03/news/9510030038_1_wallace-gator-bradley-candidates-taxes

Some of Les Golden's animal welfare and environmental activism activities are described at http://www.huffingtonpost.com/2012/05/07/les-golden-oak-park-dog-a_n_1497423.html, http://en.wikisage.org/wiki/The_Horrors_of_Zoos_and_Horse_Racing. http://en.wikisage.org/wiki/CARE_Party_of_Oak_Park

The contributions of Harvey Dubner to blackjack theory, the preliminary version of which appears in Appendix IV, are provided at http://en.wikisage.org/wiki/Harvey_Dubner

Bibliography

AFIPS Conference Proceedings (American Federation of Information Processing Societies) (1963) 1963 Fall Joint Computer Conference, Spartan Books, Baltimore, Maryland, volume 24., https://www.computer.org/csdl/proceedings/afips/1963/5063/00/506300fm.pdf.

Baldwin, Roger, Wilbert Cantey, Herbert Maisel, and James McDermott (1956) The Optimum Strategy in Blackjack, *Journal of the American Statistical Association*, 51, 429–439.

Baur, Gene (2008) *Farm Sanctuary: Changing Hearts and Minds About Animals and Food*, Simon and Schuster, New York.

Boxall, Bettina, (1999) Fate of Mapes Hotel Might Be Sealed, *Los Angeles Times*, October 11.

Blunt, Jerry (1967) *Stage Dialects*, Chandler Publishing Company, New York.

Brenner, Reuven, Gabrielle A. Brenner, and Aaron Brown (2008) *A World of Chance*: *Betting on Religion, Games, Wall Street*, Cambridge University Press, Cambridge.

Brown, Aaron (2007) *The Poker Face of Wall Street*, John Wiley & Sons, New York.

Brown, Aaron (2011) *Red-Blooded Risk*: *The Secret History of WallStreet*, John Wiley & Sons, New York.

Call for papers issued by Fall Computer Conference (2013) 1963 Fall Joint Computer Conference, *Electrical Engineering*, p. 51A, April.

Computer Clips Vegas Casino at Blackjack, *Los Angeles Times*, p. 1, November 15, 1963.

Dubner, Harriett (2017a) personal communications, January and February.

Dubner, Robert (2017b) personal communications, January and February. edwardothorp.com/id1.html; en.wikisage.org/wiki/Blackjack_card_counting_systems

Ethier, Stewart N. (2010) *The Doctrine of Chances*: *Probabilistic Aspects of Gambling*, Springer, New York.

Ethier, Stewart (2016, 2017) personal communications.

© Springer International Publishing AG 2017
L.M. Golden, *Never Split Tens!*, DOI 10.1007/978-3-319-63486-9

Ethier, S.N. and D. A. Levin, (2005) On the fundamental theorem of card counting, with application to the game of trente et quarante, *Advances in Applied Probability,* 37, 1, 90–107.

Golden, Les (2008) Card Counter Camouflage: Relativistic and Quantum Mechanical High-Tech Devices in Card Counting, *Gambling.com*, March/April, 27–31.

Golden, Les (2009) Hybrids Aren't Just for Sissies Anymore, *Bluff Europe*, February, 84–85.

Golden, Les (2010a) Check Out My Camouflaged ABS, *Bluff Europe*, July, 92–93.

Golden, Les (2010b) So, do you feel lucky, punk? Well, do ya'?, *Bluff Europe*, October, 88–89.

Golden, Les (2010c) Yonder Lies the Castle of my Fodder, *Bluff Europe*, November, 90–91.

Golden, Les (2010d) The Rain in Spain Falls Mainly on the Plain: Camouflage by Status, *Bluff Europe*, December, 90–91.

Golden, Leslie M. (2011) An analysis of the disadvantage to players of multiple decks in the game of twenty-one, *The Mathematical Scientist*, 32, 2, 57–69.

Griffin, P. (1976). The rate of gain in player expectation for card games characterized by sampling without replacement and an evaluation of card counting systems, In: *Gambling and Society: Interdisciplinary Studies on the Subject of Gambling,* Eadington, W. R. (ed), Thomas, Springfield, Illinois, 429–442.

Griffin, P. A. (1999). *The Theory of Blackjack*, 6th ed. Huntington Press, Las Vegas, Nevada.

Hart, Jennifer (2017) of the University of Chicago Crerar Library, personal communication (March 10).

Hicks, Jerry (1982) A Profile of Ed Thorp: Blackjack's No. 1 Guru, *Los Angeles Times*, date not available.

http://oracleofbacon.org/, (Kevin) Bacon number calculator.

http://www.colorradio.com/teen-queens.html, "Eddie My Love" singing group.

http://www.daledellutri.com/crf.html, Planetary elongation calendar.

http://www.truthbeknown.com/victims.htm, *How many people have died in the name of Christ, Christianity and Catholicism?: Victims of the Christian Faith.*

https://www.engadget.com/2013/09/18/edward-thorp-father-of-wearable-computing/

https://weeklytop40.wordpress.com/1956-all-charts/, Historical record sales.

Johnstone, Keith (1979) *Impro: Improvisation and the Theatre*, Faber and Faber, London.

Kahn, Joseph P. (2008) Legendary Blackjack Analysts Alive But Still Widely Unknown, *Boston Globe*, February 22; reprinted in *The Tech Online Edition*, 128, 6, February 22, 2008. http://tech.mit.edu/V128/N6/blackjack.html.

Kelly, J. L. (1956) A New Interpretation of Information Rate, *IRE Transactions on Information Theory,* IT-2, 3, September; *Bell System Technical Journal*, 35, 917–926.

Kimmel, Tim (2012) personal communication, January 12.

Kurtzman, Selma (2017) personal communication, January 26.

Maisel, Herbert (2016) personal communication, April 7.

O'Neil, Paul (1964) The Professor Who Breaks the Bank, *LIFE*, March 27, 80–91.

Patterson, Jerry (2008) The Forgottten Man of Blackjack —Harvey Dubner, http://www.readybetgo.com/blackjack/history/history-card-counting-690.html

Patterson, Scott (2010) The Quants: How a New Breed of Math Whizzes Conquered Wall Street and Nearly Destroyed It, Crown Business, New York.

Poundstone, William (2005) *FORTUNE'S FORMULA: The Untold Story of the Scientific Betting System That Beat the Casinos and Wall Street*, Hill & Wang/Farrar, Straus & Giroux, New York.

Program of the 1963 Fall Joint Computer Conference (1963) *Electrical Engineering*, p. 19A-21A, September, http://ieeexplore.ieee.org/stamp/stamp.jsp?arnumber=6539277.

Ritzenthaler, Bonnie (Mrs. Allan) Wilson (2017) personal communications, March 16–18.

Ruchman, Peter (2001a) Not Fade Away: An Appreciation of Julian Braun (1929–2000), Blackjack Forum, www.blackjackforumonline.com/content/Brauntribute3.htm, from http://www.casinogaming.com/columnists/blackjack/, February 25, March 4, March 11, March 18.

Ruchman, Peter (2001b) Thorp Steps Up to the Plate, April 1, http://www.casinogaming.com/columnists/blackjack/. (Note: The links to casinogaming.com are as of 2017 redirected to the *Las Vegas Review-Journal* newspaper. They have chosen to delete Mr. Ruchman's informative and historically significant articles, but the author have been told they are planning to rectify this act.)

Schaffer, Allan, (2005) Jess Marcum, Mathematical Genius, and the History of Card Counting, Blackjack Forum, www.blackjackforumonline.com/content/JessMarcumEarlyDaysofCardCounting.htm.

Shapiro, Sherwin (2016) personal communication.

Shinhoster Lamb, Yvonne (2008) Mathematician Co-Authored Guide to Winning at Blackjack, *Washington Post*, July 6.

Snyder, Arnold (2005), History of Blackjack: Julian Braun, Blackjack Pioneer, Blackjack Forum, www.blackjackforumonline.com/content/braunint.htm.

Snyder, Arnold (2012), *The Big Book of Blackjack*, Cardoza Publishing, Las Vegas.

Sonner, Scott (2000) Historic Mapes Hotel demolished after prolonged legal wrangling, *Las Vegas Sun*, January 31.

Thorp, E. O. (1961) A favorable strategy for twenty-one, *Proceedings of the National Academy of Sciences* 47, 110–112.

Thorp, Edward O. (1966) *Beat the Dealer: A Winning Strategy for the Game of Twenty-One*, Random House, New York.

Thorp, Edward O. (2011) personal communication, July 6.

Thorp, Edward O. (2017) personal communications, March 10, June 9.

Tupac, James D. (1963) Proceedings of the November 12–14, 1963, fall joint computer conference, Reviewers, Panelists, and Session Chairmen, ACM Digital Library, p. 641, http://dl.acm.org/citation.cfm?id=1463822&picked=prox&cfid=73716 6510&cftoken=40455784; http://portalparts.acm.org/1470000/1463822/bm/ backmatter.pdf?ip=67.243.84.66&CFID=900169367&CFTOKEN=95779162.

To Tell the Truth (1964.) Edward Thorp on "To Tell the Truth," February 3, https:// www.youtube.com/watch?v=hPIW-OJugG4.

Wilson, Allan N. (1965) *The Casino Gambler's Guide*, Harper and Row, New York.

— Jessie Beck www.beachbumtrolley.com/images/jessiebeckstory.pdf, from *Reno Evening Gazette*, March 31, 1971; *Nevada State Journal*, March 10, 1978; *Nevada State Journal*, July 18, 1987.

Index

© Springer International Publishing AG 2017
L.M. Golden, *Never Split Tens!*, DOI 10.1007/978-3-319-63486-9